LAB MANUAL
for

The 80x86 IBM PC
and Compatible Computers
(Volumes I & II)

Assembly Language, Design, and Interfacing

Second Edition

Muhammad Ali Mazidi
Janice Gillispie Mazidi

Prentice Hall
Upper Saddle River, New Jersey   Columbus, Ohio

Editor: Charles E. Stewart, Jr.
Production Editor: Alexandrina Benedicto Wolf
Cover Design Coordinator: Karrie M. Converse
Production Manager: Pamela D. Bennett
Marketing Manager: Debbie Yarnell

This book was set in Times Roman by Janice Mazidi and was printed and bound by Courier/Kendallville. The cover was printed by Phoenix Color Corp.

Printed in the United States of America

10 9 8 7 6 5 4 3 2

ISBN: 0-13-759457-7

Prentice-Hall International (UK) Limited, *London*
Prentice-Hall of Australia Pty. Limited, *Sydney*
Prentice-Hall of Canada, Inc., *Toronto*
Prentice-Hall Hispanoamericana, S.A., *Mexico*
Prentice-Hall of India Private Limited, *New Delhi*
Prentice-Hall of Japan, Inc., *Tokyo*
Editora Prentice-Hall do Brasil, Ltda., *Rio de Janeiro*

CONTENTS

PREFACE

This lab manual is designed to be used with *The 80x86 IBM PC & Compatible Computers Volumes I and II: Assembly Language, Design and Interfacing*. It is organized into five sections.

The lab assignments in Section 1 correspond sequentially with the chapters in the book. The following lists the labs beside their corresponding chapter. Chapter 1 (Labs 1, 2); Chapter 2 (Labs 3, 4, 5, 6, 7, 8); Chapter 3 (Labs 9, 10, 11, 12); Chapter 4 (Lab 13); Chapter 5 (Labs 14, 15); Chapter 6 (Labs 16, 17, 18, 19); Chapter 7 (Labs 20, 21, 22, 23); Chapter 8 (Lab 24). Lab 5 and Lab 6 provide tutorials on the use of Microsoft's Codeview and Borland's TurboDebugger, and are optional. The main purpose of lab assignments in Section 1 is to provide students a means to practice the concepts and techniques discussed throughout the textbook in the context of applications. These applications are straightforward and do not require a complicated algorithm. However, this is not the case for the labs in Section 2. To be able to perform the labs in Section 2 requires not only in-depth knowledge of the concepts and techniques covered in the textbook, but also requires an ability to formulate the needed algorithm, to code, and to test the labs. For each lab in Section 2, we have provided some hints at an algorithm; however, in some cases the solution will require further analysis by the student.

Section 3 of this lab manual provides many labs for system programming. Section 4 deals with PC interfacing via the expansion slot buses. In Section 5, the laboratory assignments involve interfacing via the PC's parallel port as well as serial ports. All the lab assignments in Section 3 require no special hardware.

The lab assignments in Section 4 require a PC bus extender and PC Interface Trainer. You can buy both of them (from Electronix Express, see Appendix E) or you can build them. Appendices B through D show how to wire-wrap each one.

Some of the lab assignments in Section 5 use a parallel port to do interfacing. Several labs in Sections 4 and 5 are the same except that in Section 4 the expansion slot signals are used while in Section 5 the parallel port signals are used.

Appendix A is a refresher for flowchart and pseudocode standards. Appendix B covers the basics of wire-wrapping. Appendix C shows how to design and wire-wrap a PC bus extender. The PC Interface Trainer design and wire-wrapping are discussed in Appendix D. A list of parts suppliers is given in Appendix E. Appendix F provides manufacturer's data sheets for many parts used in this manual.

About assemblers

Two of the most popular assemblers are MASM from Microsoft and TASM from Borland. There are also a number of shareware assemblers, among them A86 by Isaacson, that are available through many electronic bulletin boards. Unlike assemblers from Microsoft and Borland, many of these shareware assemblers do not support instructions of the 386 and higher microprocessors; therefore, some of the programs cannot be verified with such assemblers. Fortunately, both Microsoft and Borland provide their latest software packages at educational discounts to colleges and universities.

In the preparation of this lab manual, we received input and encouragement from the following individuals and we would like thank them all sincerely: Phil Golden, Danny Morse, Rusty Meadows, Rabah Aoufi, Willi Lowe, John Russell, Matt Nugent, Ronnie Pittman, Robert Schabel, Robert Jones, and Shahram Rohani.

Finally, we would welcome any suggestions for improvement of this lab manual. Write to us at the address below.

Microprocessor Education Group
P.O. Box 381970
Duncanville TX 75138
mmazidi@dal.devry.edu

SECTION 1

Assembly Language Programming

The labs in this section reinforce the material in Chapters 1 through 8 of the textbook. These labs are designed to enable the student to gain practical experience with Assembly language programming. First, programming with the DEBUG utility is explored. Appendix A of the textbook serves as a useful reference and tutorial for DEBUG. Then Assembly language programming is explored using assemblers such as MASM, TASM, and CodeView. Once the techniques for creating, assembling, and debugging are mastered, the student is then ready for exploring in greater depth programming of numerical data, ASCII and BCD data, interrupts, macros, graphics, signed numbers, strings, 32-bit data, and more.

LAB 1

USING DEBUG COMMANDS

OBJECTIVE

» To execute DEBUG commands to assemble, run, and debug Assembly language instructions

REFERENCES

Mazidi, Volumes I & II: Appendix A, Sections A.1 through A.4, Chapter 1, Sections 1.3 through 1.4

MATERIALS

80x86 IBM (or compatible) computer
DEBUG program

✓ ACTIVITY 1

Enter DEBUG, then use the R (register) command to examine the contents of all the registers of the CPU. See Worksheet question 1.

✓ ACTIVITY 2   *628-45-6789*

Use the A (assemble) command to move each digit of your ID (or Social Security) number to an 8-bit register. Use INT 3 as the last instruction in all programs in this lab to stop execution. *Note:* When using the A command, do not use an offset address less than 100. See Worksheet question 2.

✓ ACTIVITY 3

Use the U (unassemble) command to look at the code you entered in Activity 2. See Worksheet question 3.

✓ ACTIVITY 4

Use the T (trace) command to trace the program and observe changes in the registers after the execution of each instruction.

ACTIVITY 5

Enter Assembly language instructions (using the A command) to add the 8-bit registers to the register of your choice (such as AL, AH, BL). Begin the code at the offset address where INT 3 is located, to overwrite the INT 3. Do not forget to use INT 3 as the last instruction.

ACTIVITY 6

Trace the code entered in Activity 5 and analyze how the register changes. See Worksheet question 4.

ACTIVITY 7

Use the U (unassemble) command to unassemble the entire program of Activities 2 and 5, then write down the physical and logical addresses and memory contents of each location (similar to page 29 in the textbook). *Reminder:* The code segment (CS) register value in your computer will be different from the one in the book. See Worksheet question 5.

ACTIVITY 8

Part A. Enter the following code at offset 150H. The code should move the first digit of your ID number into register AL and then add seven more digits, as shown below.

```
MOV AL,4
ADD AL,6
ADD AL,2
...
INT 3
```

Part B. Trace the program above and analyze the changes in register AL. See Worksheet question 6.

ACTIVITY 9

Part A. Using the E (enter) command, put each digit of your ID or Social Security number in a memory location starting at offset address 300 (or an offset address of your choice). Verify that the data was entered correctly by using the D (dump) command. Use the A (assemble) command to write a program that adds the numbers stored in those locations (similar to the program on the bottom of page 29 in the textbook) and stores the result at offset address 320.

Part B. Using the D (dump) command, first examine the contents of offset address 320, then using the E (enter) command, put value 0 at that location. Then using the G (go) command, run the program. Examine the memory contents before and after the run and analyze the result. Make sure you have INT 3 as the last instruction before running the program.

Part C. Trace the program and analyze the changes.

Part D. Using the U (unassemble) and D (dump) commands, look at both the code and data sections of the program (U for code and D for data), then show the logical and physical addresses and their contents for both data and code (see page 29 in the textbook). See Worksheet question 7.

1. When DEBUG is entered, the general-purpose registers are set to what value? When you entered DEBUG for Activity 1, what values did the segment registers have for your PC?

AX = 0000   DS = OC6E
BX = 0000   ES = OC6E
CX = 0000   SS = OC6E
DX = 0000   CS = OC6E

2. Assuming that in Activity 2, all 8-bit general-purpose registers were used for 8 digits of your Social Security number, which register did you choose for the 9th digit? Why was that register chosen?

BP because it was set at 0000

3. Examine the unassembled code for Activity 3. Is the machine code the same for each MOV command? In other words, is the machine code to move a value into register AL the same as the machine code to move a value into AH? *no*

AL = B0
AH = B4

4. Attach the trace of your program in Activity 6. This can be done by using Shift-PrintScreen after it is on the screen, or by using Ctrl-PrintScreen before entering the trace command. Shift-PrintScreen prints what is on your screen. After Ctrl-PrintScreen is entered, anything that appears on the screen will be sent to the printer until Ctrl-PrintScreen is entered again. On your printout, circle the register used to accumulate the sum as it changes, and place a box around the final sum.

5. Examine the trace of your program in Activity 7. How many bytes are needed for the machine code of a MOV instruction? For an ADD instruction?

6. Based on your observations of machine code, write down what you think would be the machine code for "~~MOV~~ AL,23H".
   ADD

   04 23

7. Attach the trace of your program in Activity 9.

LAB 2

EXAMINING THE FLAG BITS

OBJECTIVES

» To use DEBUG to examine and alter the contents of the flag register
» To examine how Assembly language instructions affect and are affected by the flag register

REFERENCES

Mazidi, Volumes I & II: Appendix A, Sections A.1 through A.4, A.6
Chapter 1, Sections 1.3 through 1.5

MATERIALS

80x86 IBM (or compatible) computer
DEBUG program

ACTIVITY 1

Use DEBUG to add following hex numbers: 1222H, 1333H, 1191H, 32F1H. *Reminder:* DEBUG assumes that the numbers are in hex (no H is needed).

```
MOV  AX,0
ADD  AX,1222
....    ...
```

Trace the program and analyze the ZF, AC, SF, PF, and CF flag bits. See Worksheet question 1.

ACTIVITY 2

Rewrite Activity 1 using a loop. See Worksheet question 2.

ACTIVITY 3

Using the A (assemble) command, write a program that adds the digits of your ID (or Social Security) number, then trace and analyze the flag bits SF, PF, CF, AC, and ZF. In this program do not use a loop.

ACTIVITY 4

Repeat Activity 3, using a loop. See Worksheet question 3.

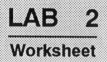

Name: _____

Date: _____

Class: _____

1. What was the final value in AX after running Activity 1? List the status of flag bits ZF, AF, SF, PF, and CF, then briefly state what each condition code indicates. Show the logical address and physical address of each instruction.

   $AX = 69D7$

2. Observe the changes in the flag bits as the last iteration of your loop in Activity 2 is completed. Which flag(s) determine whether the loop continues or finishes for the conditional jump you used in the program? Which flag value(s) will cause the jump to occur? Which flag value(s) will cause the jump not to occur?

3. What is the sum of the digits of your Social Security number? What was the status of flag bits ZF, AF, SF, PF, and CF when the final digit was added in Activity 4? Describe what those states of the flag bits mean.

4. Fill in the following table by listing the condition codes used in the DEBUG register dump for the indicated flag when the flag = 0 in the first column and the condition codes when the flag = 1 in the second column.

Flag	Reset (= 0) Value	Set (= 1) Value
SF, sign flag		
ZF, zero flag	PL	NG
OF, overflow flag	NV	OV
PF, parity flag	PE	PO
CF, carry flag	NC	CY
DF, direction flag	UP	DN
IF, interrupt flag	DI	EI
AF, auxiliary carry flag	NA	AC

5. Enter DEBUG, then display the registers with the R command. Demonstrate that you can change the status of the flag bits by changing the values of 3 of the flag bits, then redisplay the flag register. Attach a trace.

LAB 3

USING AN ASSEMBLER

OBJECTIVES

» To code an Assembly language program with code, data, and stack seg-
ments
» To create an executable ".exe" file by assembling and linking the code
» To use DEBUG with an ".exe" file

REFERENCES

Mazidi, Volumes I & II: Chapter 2, Sections 2.1 through 2.2.

MATERIALS

80x86 IBM (or compatible) computer
MASM (or compatible) assembler
DEBUG program

ACTIVITY 1

Part A. Use the assembler of your choice to write, assemble, and link
a program that adds all the digits of your ID (or Social Security) number. Use
a loop and the register indirect addressing mode in writing this program. You
must include data and code segments.

Part B. Using DEBUG, load the ".exe" file created in Activity 1 and
then unassemble it (see Program 2-2 in the textbook) to see the value assigned
to the data segment register (DS) by DOS. Using the D command of DEBUG,
dump your data in the data segment to examine the placing of your data by
the assembler.

Part C. Trace the program and analyze the changes in the registers
and flag bits SF, ZF, AC, PF, and CF. Highlight the accessing of data from
the data segment. See Worksheet questions 2 and 3.

ACTIVITY 2

Repeat Activity 1, this time adding 4 words of data. Use the data
words given in LAB 2 Activity 1: 1222H, 1333H, 1191H, and 32F1H. Save
this program because it will be used again in Lab 8.

ACTIVITY 3

Write a program to move a block of data one byte at a time.

ACTIVITY 4

Repeat Activity 3, this time moving a block of data one word (two bytes) at time. Use the data words given in Activity 3. See Worksheet question 4. Save this program because it will be used again in Lab 8.

Programming Tip:

In the following code, part (a) is a loop to add several bytes pointed at by SI whereas part (b) is a loop which adds several words.

(a)

```
ADD_LP:    ADD  AL,[SI]
           INC SI
           DEC CX
           JNZ ADD_LP
```

(b)

```
ADD_LP:    ADD  AX,[SI]
           INC SI
           INC SI
           DEC CX
           JNZ ADD_LP
```

Notice in the ADD instructions that neither BYTE PTR nor WORD PTR was used to indicate whether SI should point to a word or a byte. How does the CPU know whether to get a word or a byte? The answer is that the CPU looks at the destination operand and assumes that SI will point to a byte if the destination operand is a byte and it will assume that the operand is a word if the destination operand is a word.

---

1. For the following program, place a box around the Assembly language instructions. For each directive, write in the comment area what the directive does.

```
STSEG      SEGMENT
           DB     64 DUP (?)
STSEG      ENDS
;----------------------------
DTSEG      SEGMENT
DATA1      DB     52H
DATA2      DB     29H
SUM        DB     ?
DTSEG      ENDS
;----------------------------
CDSEG      SEGMENT
MAIN       PROC FAR
           ASSUME CS:CDSEG,DS:DTSEG,SS:STSEG
           MOV   AX,DTSEG
           MOV   DS,AX
           MOV   AL,DATA1
           MOV   BL,DATA2
           ADD   AL,BL
           MOV   SUM,AL
           MOV   AH,4CH
           INT   21H
MAIN       ENDP
CDSEG      ENDS
           END   MAIN
```

2. Attach the list file generated in Activity 1. Underscore the data directives
   used in your program.

3. Attach the list file generated in Activity 1, and the DEBUG dump of the
   data. Compare the listing of the data in the ".lst" file with the dump of the
   data in DEBUG. Circle the ID (or Social Security) number data item on
   both the list file and the data dump. Place a box around the data item used
   for the sum of the digits on both the list file and the data dump.

4. Compare the dump of the data segment in DEBUG before and after the program is executed in Activity 4. Circle all changes in the data segment
   caused by your program.

LAB 4

EXAMINING DATA TYPES

OBJECTIVES

» To code Assembly language directives to set aside memory locations in the data segment and give them initial values

» To examine the data segment of a program in DEBUG

REFERENCES

Mazidi, Volumes I & II: Chapter 2, Sections 2.1 through 2.2, 2.5

MATERIALS

80x86 IBM (or compatible) computer
MASM (or compatible) assembler
DEBUG program

ACTIVITY 1

Use the assembler of your choice to code a program similar to the shell program shown in the textbook in Figure 2-2. For the data segment, use the data given in Problem 15 of Chapter 2. Assemble and link the program.

ACTIVITY 2

Use DEBUG to examine the DS value assigned by DOS in your computer then dump the data. Show the exact logical and physical address where each byte of data is located.

ACTIVITY 3

Repeat Activity 2 on a different computer to see if the DS value is the same as in Activity 2. See Worksheet question 1.

ACTIVITY 4

Observe how the following data items are placed in memory by placing the following DATA segment in a program, assembling and linking the program, then dumping the data in DEBUG. See Worksheet question 2.

```
        DSEG SEGMENT
            ORG  10H
DATA1 DB    '09876543'
            ORG  20H
DATA2 DD    098765432H
            ORG  30H
DATA3 DW    1000100101100100B,8964H
            ORG  40H
DATA4 DB    4 DUP ('-', 2 DUP ('*'))
        DSEG ENDS
```

ACTIVITY 5

For the following DATA segment, anticipate how the data will be stored in memory. Write out the relevant data as it would be dumped in DEBUG. Then place this data segment in a program shell, assemble, link, and DEBUG it to see if the data was stored as you anticipated. See Worksheet question 3.

```
        DSEG SEGMENT
            ORG  10H
DATA1 DB    '1909 HILLTOP ROAD'
DATA2 DB    '1234'
DATA3 DW    1234H
DATA4 DW    1234
            ORG  40H
DATA5 DD    12345678
DATA6 DD    12345678H
        DSEG ENDS
```

Programming Tip:

When writing large programs, it can be tedious to keep track of where each variable is defined and updated in the program. MASM provides a helpful tool in keeping track of your variables called the cross-reference file. To generate the cross-reference file, first create a ".crf" file when assembling:

```
C>MASM PROG1.ASM <CR>

Microsoft (R) Macro Assembler  Version 5.10
Copyright (C) Microsoft Corp 1981, 1988.  All rights reserved.

Object filename [PROG1.OBJ]:<CR>
Source listing  [NUL.LST]:PROG1.LST <CR>
Cross-reference [NUL.CRF]: PROG1.CRF<CR>
```

Then generate the cross-reference listing by using the CREF program as follows:

```
C>CREF PROG1;
```

The above command creates a file named PROG1.REF which lists each name, where it was defined, and each place it was referenced by line number. These line numbers correspond to the line numbers on the ".lst" file. By printing PROG1.REF and PROG1.LST, you can easily track down any variable

---

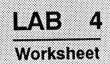
1. For Activities 2 and 3, did you have the same value for the DS register? State your conclusions about DOS memory management in terms of mapping the logical addresses of program segments into physical memory.

2. Based upon your observations in Activity 4, answer the following questions.

   (a) ASCII data should be stored with which directive?

   (b) Describe how and why DATA1 and DATA2 were stored differently.

   (c) Describe the little endian storage convention as demonstrated by the placement of data for DATA3.

   (d) Is the following legal?   4 DUP ('-', 2 DUP '*')
       If not, indicate what changes should be made to make it legal.

3. Compare the anticipated versus the actual storage of the data items in Activity 5. Were any of the items stored differently than you anticipated? If so, what did you learn from this exercise?

LAB 5

USING MICROSOFT CODEVIEW

OBJECTIVE

» To debug a program using Microsoft's CodeView debugger

REFERENCES

Mazidi, Volumes I & II: Chapter 2, Section 2.3

MATERIALS

80x86 IBM (or compatible) computer
MASM (or compatible) assembler
Microsoft's CodeView

In this lab we show some of the basic commands of CodeView. A complete description of CodeView can be found in Microsoft's manuals. The information provided here is enough to allow you to debug simple programs.

Setting up a program for CodeView

In order to have symbolic information available in CodeView, the Assembly language program must have been assembled with the /ZI option:

C>MASM PROG1.ASM /ZI

and linked with the /CO option:

C>LINK PROG1.OBJ /CO

Using these options will enable you to look at the source code in CodeView. CodeView can be used with files created without the above assemble/link options.

Entering CodeView

The following shows how to enter CodeView to debug PROG1.EXE

C>CV PROG1

CodeView can also be used for any ".exe" file generated with C, Pascal, or other compilers.

```
4B35:0000 B8344B        MOV     AX,4B34              AX=      6DEE
4B35:0003 8ED8          MOV     DS,AX                BX=      0000
4B35:0005 B90500        MOV     CX,0005              CX=      0000
4B35:0008 BB0000        MOV     BX,0000              DX=      0000
4B35:000B B000          MOV     AL,00                SP=      0200
4B35:000D 0207          ADD     AL,Byte Ptr [BX]     BP=      0000
4B35:000F 43            INC     BX                   SI=      0000
4B35:0010 49            DEC     CX                   DI=      0000
4B35:0011 75FA          JNZ     000D                 DS=....4836
4B35:0013 A20500        MOV     Byte Ptr [0005],AL   ES=....4823
4B35:0016 B44C          MOV     AH,4C                FS=....0000
B35:0018  CD21          INT     21                   GS=....0000
4B35:001A 5A            POP     DX                   SS=....0837
4B35:001B 17            POP     SS                   CS=....4B35
4B35:001C 2C06          SUB     AL,06                IP=0000002C
4B35:001E A31740        MOV     Word Ptr [4017],AX
4B35:0021 06            PUSH    ES                      NV UP
4B35:0022 E717          OUT     17,AX                   EI NG
                                                        NZ AC
>t                                                      PE NC
>t
>t                                                      DS:0004
>                                                       00000000
```

Sample CodeView Screen

The box above shows a sample CodeView screen. This is the screen produced when Program 2-1 in the textbook was run in CodeView. The highlighted line shows the next instruction to be executed. Notice the prompt ">" that appears at the bottom of the CodeView screen. The following is a list of some of the more commonly used commands that can be typed in at this prompt. Many of these commands can be initiated from menus with the use of a mouse, if your system has one.

r	displays register values at bottom of screen, similar to DEBUG
a	assemble instructions into program
db	display bytes of memory
dw	display words of memory
eb	enter bytes in memory
g	go (execute) until breakpoint
p	execute current source line (skips routines and interrupts)
q	quit
t	trace (execute) current source line (does not skip routines and interrupts)
w	watch (variable or expression)

Following are some frequently used Function Keys in CodeView

F2	toggle register display on and off
F3	toggle between Assembly and source listing
F4	toggle between output and program screen

F5	execute until breakpoint
F6	toggle cursor between program and command prompt ">"
F8	single-step (including CALL and INT subroutines)
F9	manage breakpoints
F10	single-step (excluding CALL and INT subroutines)

Displaying and entering data

Data can be displayed in byte (DB) or word (DW) format. These two commands display the data in hex. CodeView also allows you to display the data in decimal with the DU command. A number can follow the D command to tell CodeView what offset to begin displaying at. Repeatedly entering a D command will go through the data segment, chunk by chunk.

>DB 0 36	displays data segment from offset 0 for 36 bytes
>DU 0 36	displays decimal equivalents of bytes from offset 0 to 36
>DW 0 36	displays words from offset 0 to 36

Notice that the numbers entered above are in decimal, not hex. For example, "DB 0 36" will dump 36 bytes, which is equivalent to 24 hex. Data can also be entered as hex bytes (EB) or words (EW), or as decimal integers (EI). The number following the E command indicates at what offset to begin entering the data. A few examples are:

>EB 16 0 0 0	enters three 0 bytes starting at offset 16
>EW 16 0 0 0	enters three 0 words starting at offset 16
>EA 16 32 32 32	enters three ASCII blanks at offset 16
>EI 0 -10 10 20	enters integers (-10, 10, 20) at offset 0

Watch commands

The watch commands are very useful tools in CodeView. They can be typed in as commands on the command line or can be set up by using the "Watch" menu at the top of the CodeView screen. Watch commands set up a watch box at the top of the screen where you can watch data change as the program executes. Some examples will help show how watch commands can be used.

>WB DS:0 L16	this will watch the data segment from offset 0 for 16 bytes
>W? DATA_1	this will watch data item named DATA_1
>WP? SUM > 0	this will watch until data item SUM is greater than 0

The above three commands show some of the most useful aspects of the watch command. The first command showed the watch command being used to dump specific offset values of the data segment. The second command sets up a watch for a specific variable. This command will only work if the program was assembled and linked with the options shown on the first page of this lab. The third command showed how to set up breakpoints in the program. If the go ">G" command is entered after this command, the program will execute until the variable SUM is greater than zero, or until the program terminates.

Helpful hints

1. As the program executes, it scrolls upwards off the screen. At times you may wish to look at the code that has already scrolled off the screen. In this case, the U command could be used. For example, if you use the U0 command to unassemble starting at offset 0, this will get you back to the beginning of the program.

2. If you try to display the data segment before it has been set up in the program "MOV DS,AX", you may have some difficulty. Therefore, go ahead and single-step through the program past those initial housekeeping instructions before trying to look at the data segment.

ACTIVITY 1

Prepare Program 2-2 in the textbook for debugging by assembling and linking the program with the options shown above. Enter CodeView as shown in this lab, and hit F2 to display the registers. Now at the ">" prompt, use the r command to display the register values at the bottom of the screen. Experiment with the r (register), d (data), and e (enter) commands. These three commands (and many others) work similarly to the way they work in DEBUG.

ACTIVITY 2

Single-step through the program, then examine the result data in the data segment. See Worksheet question 1.

ACTIVITY 3

Change the data to be added in the data segment by using the e command to enter the 4 words: 1234, 0F24, 032B, 1122. Then execute the program with the new data and verify that the correct result was obtained. See Worksheet question 2.

1. Use Ctrl-PrtScreen to print the screen at the point in your trace where the result is loaded into SUM. List the commands you used to verify that the correct data was placed in the data segment.

2. List the series of commands you used to enter the data, run the program, and verify the results.

3. Give a summary of the features of CodeView which you found most useful. Compare it with another debugging tool such as DEBUG.

LAB 6

USING BORLAND TURBO DEBUGGER

OBJECTIVE

» To debug a program using Borland's Turbo Debugger program

REFERENCES

Mazidi, Volumes I & II: Chapter 2, Section 2.3

MATERIALS

80x86 IBM (or compatible) computer
DEBUG program
Borland's Turbo Debugger (version 3.0 used here)

In this lab we show some of the basic commands of Turbo Debugger. A complete description of Turbo Debugger can be found in Borland's manuals. The information provided here is enough to allow you to debug simple programs.

Setting up a program for Turbo Debugger

The Assembly language program should be assembled with the /ZI option:

C>TASM /ZI TEST.ASM
and linked with the /V option:
C>TLINK /V TEST1.OBJ

Note that when using Borland's Turbo Assembler and Link programs, the commands are TASM and TLINK, not MASM and LINK. Using the options shown above will enable you to look at the source code in Turbo Debugger, but the debugger can be used with files created without the above assemble/link options.

Entering Turbo Debugger

The following shows how to enter Turbo Debugger to debug TEST.EXE

C>TD TEST

```
  File Edit  View  Run  Breakpoints  Data  Options  Window  Help        READY +-
 [_] Module: test File: test.asm 15                              1-[][]-+
    PAGE      60,132
    TITLE     PROG2-1 (EXE) PURPOSE: ADDS 5 BYTES OF DATA
    STSEG     SEGMENT
              DB    32 DUP (?)
    STSEG     ENDS
    DTSEG     SEGMENT
    DATA_IN   DB    25H,12H,15H,1FH,2BH
    SUM       DB    ?
    DTSEG     ENDS
    CDSEG     SEGMENT
    MAIN      PROC       FAR
              ASSUME CS:CDSEG,DS:DTSEG,SS:STSEG
              MOV   AX,DTSEG
              MOV   DS,AX
              MOV   CX,05           ;set up loop counter CX=5
              MOV   BX,OFFSET DATA_IN    ;set up data pointer BX

       Watches

 +
  F1-Help F2-Bkpt F3-Mod F4-Here F5-Zoom F6-Next F7-Trace F8-Step F9-Run F10-Menu
```

Sample Borland Turbo Debugger Screen

Getting around in Turbo Debugger

The above shows roughly how the Borland Turbo Debugger screen looks. Each command at the top of the screen (File, Edit, View, etc.) represents a category of commands which are accessed by a drop down menu. If you are using a mouse, simply click on the command to get the drop down menu, then click on the desired item in the drop down menu. If you do not have a mouse, you select commands by holding down the Alt key while pressing the letter of the command that is highlighted in red. For example, to select **File**, press Alt-F, or to select **View**, press Alt-V. To exit a drop down menu, hit the Esc key. The bottom line of the screen shows some commonly used commands which can be accessed by function keys. When a command is selected, either by a mouse or by the Alt key, a line at the bottom of the screen tells you what that command will do. F10 toggles the cursor between the program and the command line at the top of the screen

The View menu

This drop down menu provides several windows into the system. For example, if you select the **Register** option, a window will appear which displays the contents of the registers. If you select the **Dump** option, a memory dump of the data segment will appear. Other useful views are the **Variables** window and the **Breakpoints** window. The **Variables** command shows the value of a variable. To set up a watch on a variable, click on the variable in the program, then hit Ctrl-F7. Once the windows are displayed on the screen, you can move them around on the screen by using the mouse. To exit the window, double click on the little green box in the top left corner.

The Run menu

This menu provides several run options. The **Run** option (also available by hitting F9) runs the program until a breakpoint is reached or until the end of the program. **Go to cursor** allows you to set up a spontaneous breakpoint by executing the program until the line with the current cursor position. F7 can be used to single-step through the program. F8 will single-step through the program, but will skip over procedure and macro calls. If you want to start the program from the beginning, hit Ctrl-F2 and the program pointer will be set to the beginning of the program. If your program has screen output, you can view it by toggling Alt-F5.

The Breakpoints menu

The **Toggle** command (also available by hitting F2) allows you to toggle a breakpoint on and off the line of code at the current cursor position. You can set up as many breakpoints as you need by moving the cursor to the desired line (by the mouse or the up/down arrow keys), then hitting F2. To remove all breakpoints, select the "**Delete** all" option from the **Breakpoint** menu.

The Data menu

The **Inspect** command in the **Data** menu is useful for watching variables. After you select **Inspect**, you will be prompted for the variable names you wish to inspect. Be aware that this command may be case sensitive; type the variables in the same case as they appear in the program.

The CPU window

This extremely powerful window is available in the View drop down menu. It has so many features that it is discussed here under a separate heading. In fact, this option allows you such access to the CPU that it is scary! You can examine and alter your code as well as the data segment. The CPU window is composed of five panes, containing the following: unassembled code with source, registers, flags, data dump, stack dump. Within any window (or pane), hitting Alt-F10 pops up a menu allowing you to make changes to the window or pane.

In the Code pane of the CPU window, both the Assembly language instructions and the machine code are displayed. While in this pane, you can use the F7 or F8 single-step commands, as well as the breakpoint and run commands from the menus. In the Alt-F10 local menu, the assemble command will let you assemble an instruction, replacing the one at the current cursor position.

In the Register pane of the CPU window, the registers are all displayed in hex. The Alt-F10 local menu can be used to increment, decrement, reset, or change the value of any register. Register values can also be changed by highlighting them and typing in a new value.

The flags pane also allows you to change the value of the flag bits which are indicated by the first letter of their name (e.g., c = carry, s = sign).

The data pane of the CPU window dumps memory from any of the four segments of your program. You can alter the data by highlighting it and then typing in the new data. The Alt-F10 local menu provides many options such as searching for a character string, setting a block of memory to a value, changing the display format to byte, word, or other options.

ACTIVITY 1

Prepare Program 2-2 in the textbook for debugging by assembling and linking the program with the options shown in this lab. Enter Turbo Debugger, use the View menu to set up a register window, and a CPU window. Single-step through the program, then examine the result data in the data segment. See Worksheet question 1.

ACTIVITY 2

Change the data to be added in the data segment to these 4 words: 1234, 0F24, 032B, 1122. Then execute the program with the new data and verify that the correct result was obtained. See Worksheet question 2.

ACTIVITY 3

Experiment with the CPU window by changing register values, data segment values, and program instructions.

---

1. Use Ctrl-PrtScreen to print the screen at the point in your trace where the result is loaded into SUM in Activity 1.

2. Describe how you entered the data, ran the program, and verified the results for Activity 2.

3. Give a summary of the features of Turbo Debugger which you found most useful. Compare it with another debugging tool such as DEBUG or CodeView.

LAB 7

USING THE SIMPLIFIED SEGMENT DEFINITION

OBJECTIVE

» To code Assembly language directives that define segments with the simplified format

REFERENCES

Mazidi, Volumes I & II: Chapter 2, Section 2.6

MATERIALS

80x86 IBM (or compatible) computer
MASM (or compatible) assembler
DEBUG program

SIMPLIFIED SEGMENT DEFINITION

When the simplified segment definition is used, the offset addresses in the data segment are overruled. For example, notice the use of offset addresses below.

```
                .MODEL SMALL
                .STACK      64H
                .DATA
DATA_1          DB     4,6,2,3,7,8,6,9
                ORG    10H
SUM_B           DB     ?
                ORG    18H
DATA_2          DW     1222H,1333H,1191H,7869H
                ORG    28H
SUM_W   DW     ?
                .CODE
                MOV    AX,@DATA
                MOV    DS,AX
                ...
                ...
                ...
```

Using DEBUG, you can see that the offset address of 10H for SUM_B is not honored. The same is true for the other offset addresses 18H and 28H. This is because MASM is trying to be efficient in allocating memory. Therefore, in using the new segment definition it is strongly recommended not to assign an offset addresses to a variable. This is also recommended for any real-life program, regardless of whether you are using the traditional segment definition or the new segment definition. The only reason we have done so in the textbook is to separate the data fields for the sake of clarity when using DEBUG.

In full segment definition, the base address for calculating an OFFSET address is the data segment address. With simplified segment definition, the base address is the DGROUP segment address for all segments associated with the group. For example, look at the following instruction:

MOV   AX,OFFSET NUM_1

If the above statement were in a program using full segment definition, the offset of NUM_1 would be relative to the beginning of the data segment. However, if the above statement were in a program using simplified segment definition, the offset of NUM_1 would be relative to the start of the DGROUP. For more about DGROUP, see the section below. Another difference between full and simplified segment definition is that some defaults differ between the two formats. For example, the default alignment is PARA in full segment definition, but for simplified segment definition with the SMALL model, it is WORD for code and data segments, PARA for stack segments. If these defaults need to be overridden, full segment definition must be used.

THE USE OF DGROUP

A *group* is a collection of segments associated with the same starting address. They reside within the same 64K segment. This is useful in situations where more than one segment should be associated with the same segment register at run time. A group named DGROUP is automatically created when using the simplified segment definition. DGROUP includes the .DATA and .STACK segments, but not the .CODE segment. DGROUP can be thought of as the automatic data group. When using the simplified segment definition, the address assigned by instructions which access memory in the data segment will be the address of DGROUP, not the address of the data segment.

ACTIVITY 1

Rewrite the programs you wrote for Lab 3, Activities 2 and 4, using the simplified segment definition. Produce ".exe" and ".lst" files for these programs. Then reassemble and link (if necessary) the full segment definition versions written for Lab 3 to produce ".exe" and ".lst" files. See Worksheet question 1.

ACTIVITY 2

Run the simplified segment versions in CodeView and DEBUG to see the placement of the data. Trace through the instructions which access variables in the data segment and note the offset addresses used. See Worksheet question 2.

1. Compare the size of the source files (generated in Activity 1) for full segment definition programs versus the simplified programs. What is the difference in terms of number of lines of code? Which version do you find easier to use and why? What are some limitations of using the simplified version?

2. In Activity 2, were the same offset addresses used in the list files for the full and simplified versions of the programs? When the programs were traced, were these offset addresses honored for the full segment version? For the simplified version? If they differed, for each program, list the offset addresses in the list file and the corresponding offset addresses actually encountered during the trace of the programs.

LAB 8

GENERATING A COM FILE

OBJECTIVE

» To rewrite Assembly language programs into COM format

REFERENCES

Mazidi, Volumes I & II: Chapter 2, Section 2.7

MATERIALS

80x86 IBM (or compatible) computer
MASM (or compatible) assembler
DEBUG program

ACTIVITY 1

Rewrite the program written in Lab 3 Activity 2 to change it into a COM file. Compare this file with the EXE file written for Lab 3 in terms of size of the files (look at the directory to see the file sizes). Also generate ".map" files for the programs. See Worksheet question 1.

ACTIVITY 2

Rewrite the program written in Lab 3 Activity 4 to change it into a COM file. Compare this file with the EXE files written for Lab 3 in terms of size of the files (look at the directory to see the file sizes). Trace through the EXE program in DEBUG, then trace through the COM program. See Worksheet question 2.

1. State the differences in the maps for the COM and EXE versions of the program. State how the program entry points and size of the code segment are affected by the "OFFSET 100H" statement before the Assembly language code in the COM version of the program.

2. Describe the placement of the code and the data within the program for EXE versus COM files. When you unassembled the code in the EXE file, what offset address did you use? For the COM file?

3. Using DEBUG, give the segment register contents of each COM file in Activities 1 and 2.

4. How would you prove that the DEBUG program is a COM file. Of course, if you do a DIR directory listing in the directory where it is located, you will see that it is a COM file. The question here is, by using DEBUG, how can you tell it is a COM file, not an EXE file?

LAB 9

ADDITION OF BYTE, MULTIBYTE, WORD, AND MULTIWORD OPERANDS

OBJECTIVE

» To code Assembly language instructions to perform unsigned addition of byte, multibyte, word, and multiword operands

REFERENCES

Mazidi, Volumes I & II: Chapter 3, Section 3.1

MATERIALS

80x86 IBM (or compatible) computer
MASM (or compatible) assembler
DEBUG program

ACTIVITY 1

Fill in the table for Worksheet question 1 by executing the indicated instructions in DEBUG and observing the results.

ACTIVITY 2

Fill in the table for Worksheet question 2 by executing the indicated instructions in DEBUG and observing the results.

ACTIVITY 3

Write a subroutine to find the total sum of many individual bytes and store the word-sized result. The choice of the number of bytes and the data is up to you but the total sum must be more than FFH.

ACTIVITY 4

Write a subroutine to find the sum of two multibyte numbers, each of 8-byte size. The addition must be done one byte at a time.

ACTIVITY 5

Write a subroutine to find the total sum of many individual words and store the doubleword-sized result. The choice of the number of words and the data is up to you but the total sum must be more than FFFFH.

ACTIVITY 6

Write a subroutine to find the sum of two multibyte numbers, each of 8-byte size. The addition must be done one word at a time.

ACTIVITY 7

Write a subroutine to subtract one multibyte (8-byte) number from another and save the result. The subtraction should be performed one byte at a time.

ACTIVITY 8

Write a subroutine to subtract one multibyte (8-byte) number from another and save the result. The subtraction should be performed one word at a time.

ACTIVITY 9

Write the main program which calls the above subroutines. Assemble it, run and analyze the data dump before and after the run. See Worksheet question 3.

1. Fill in the following table for Activity 1.

Operation	Destination	Source	Result	CF	SF	PF	AF	ZF
ADD	AL = 64H	46H						
ADD	BL = FFH	1						
ADD	AX = 1234H	1DF3H						
SUB	AL = 0H	1						
SUB	CX = F934H	F000H						
SUB	AX = 1234H	1234H						

2. Fill in the following table for Activity 2.

Operations		Destination	Source	Result	CF	SF	PF	AF	ZF
STC	ADC	AL = 64H	46H						
CLC	ADC	AX = 1234H	1DF3H						
STC	SBB	AL = 0H	1H					.	
STC	SBB	AX = F934H	F000H						
STC	ADC	BL = FFH	0						

3. For each subroutine written in Activities 3 through 8, circle and label the result of each subroutine in the DEBUG dump of the data segment.

4. When should the ADC instruction be used instead of the ADD instruction?

5. When should the SBB instruction be used instead of the SUB instruction?

LAB 10

MULTIPLICATION

OBJECTIVE

» To code Assembly language instructions to perform unsigned multiplication

REFERENCES

Mazidi, Volumes I & II: Chapter 3, Section 3.2

MATERIALS

80x86 IBM (or compatible) computer
MASM (or compatible) assembler
DEBUG program

ACTIVITY 1

Fill in the table for Worksheet question 1 by executing the indicated instructions in DEBUG and observing the results.

ACTIVITY 2

Write, assemble, run, and verify a program with the following objectives. Assume an individual's normal work week is 40 hours at $12 an hour. In addition, this individual has daily overtime pay of $18 an hour. The program should calculate the weekly pay of the individual by adding the regular and overtime pay, then store the result. See Worksheet question 2.

The daily overtime hours for the week are provided in the following manner:

```
OVER_8_HRS      DB      3,5,2,4,1
```

ACTIVITY 3

A given inventory has a total of 5 categories of products and the number of items in each category and the price per item is provided as follows:

```
ITEMS       DW   5,9,30,12,98
PRIC_PER_1 DW    2555,129,459,190,96
```

Write, run, and verify a program that finds the total inventory price. In the above data definition there are 5 items in category one and the price of each is $2555, the second category has 9 items each priced at $129, and so on.

1. Fill in the following table by coding and executing the indicated instructions in DEBUG and observing the results. For example, for the first instruction, execute the following, then observe the results.

   MOV AL,64
   MOV BL,46
   MUL BL

Instruction	AL or AX	Register 2	Result	CF	SF	PF	AF	ZF
MUL BL	AL = 64H	BL =46H						
MUL CH	AL = FFH	CH =1						
MUL BX	AX = 1234H	BX= 1DF3H						
MUL DL	AL = 0H	DL=1						
MUL CX	AX = F934H	CX=F000H						
MUL AX	AX = 1234H							

2. For Activity 2, dump the data segment before and after running the program, make a print-out of it, then circle the final result in the data segment.

3. For Activity 3, dump the data segment before and after running the program, make a print-out of it, then circle the final result in the data segment.

4. State the difference between "MUL BL" and "MUL BX" and explain how the CPU differentiates between byte × byte and word × word operations.

5. State a major limitation of MUL instructions in the 80x86 microprocessor. *Hint:* Think in terms of register usage.

6. True or False. Word × byte multiplication is really word × word multiplication. Explain your answer.

LAB 11

DIVISION

OBJECTIVE

» To use the DIV command to perform unsigned division in Assembly language programs

REFERENCES

Mazidi, Volumes I & II: Chapter 3, Section 3.2

MATERIALS

80x86 IBM (or compatible) computer
DEBUG program
MASM (or compatible) assembler

ACTIVITY 1

Fill in the table for Worksheet question 1 by executing the indicated instructions in DEBUG and observing the results.

ACTIVITY 2

Write, run, verify, and analyze the results of a program that finds the highest, the lowest, and the average grade of a class of 18 students. Use your own data.

ACTIVITY 3

Write, run, and verify a program that finds the highest, the lowest, and the average monthly paycheck of a salesperson for 12 months of the year.

ACTIVITY 4

Using the data from Lab 10 Activity 3, find the total price for each category and then find the lowest and highest total category price. In addition, find the average total price of all 7 categories.

1. Fill in the following table by coding and executing the indicated instructions in DEBUG and observing the results. For example, for the first instruction, execute the following, then observe the results.

   MOV AL,64
   MOV BL,46
   DIV BL

   For the Result column, indicate the register(s) which hold the result as well as what the result is.

Instruction	AL or AX	Register 2	Result	CF	SF	PF	AF	ZF
DIV BL	AL = 64H	BL =46H						
DIV BL	AL = FFH	BL =1						
DIV BX	AX = 1234H	BX= 1DF3H						
DIV CH	AL = 0H	CH=1						
DIV CX	AX = F934H	CX=F000H						
DIV BX	AX = 1234H	BX=1234H						

2. For Activity 2, dump the data segment before and after running the program, make a print-out of it, then circle the final result in the data segment.

3. For Activity 3, dump the data segment before and after running the program, make a print-out of it, then circle the final result in the data segment.

4. State a case where the "divide error" message would be displayed. Test your answer using DEBUG.

5. True or False. Byte/byte operations are really word/byte operations. Explain your answer.

6. State a major limitation of the DIV operation in 80x86 machines.

LAB 12

BCD & ASCII ADDITION AND SUBTRACTION

OBJECTIVES

» To code Assembly language instructions to perform addition and subtraction on packed BCD numbers
» To code Assembly language instructions to perform addition and subtraction on ASCII data

REFERENCES

Mazidi, Volumes I & II: Chapter 3, Section 3.4

MATERIALS

80x86 IBM (or compatible) computer
MASM (or compatible) assembler
DEBUG program

ACTIVITY 1

Using DEBUG, write Assembly language instructions to add two unpacked BCD numbers and store the result in BCD. Test the instructions on at least 3 sets of data.

ACTIVITY 2

Using DEBUG, write Assembly language instructions to add two unpacked BCD numbers and store the result in hex. Test on the same data you selected for Activity 1. See Worksheet question 2.

ACTIVITY 3

Using DEBUG, write Assembly language instructions to subtract two unpacked BCD numbers and store the result in BCD. Test the instructions on at least 3 sets of data.

ACTIVITY 4

Using DEBUG, write Assembly language instructions to subtract two unpacked BCD numbers and store the result in hex. Test on the same data you selected for Activity 2. See Worksheet question 3.

ACTIVITY 5

Write a subroutine to add two 10-digit ASCII numbers and store the result in ASCII. See Worksheet question 4.

ACTIVITY 6

Write a subroutine to subtract two 10-digit ASCII numbers and store the result in ASCII. See Worksheet question 5.

ACTIVITY 7

Write a subroutine to add two 10-digit BCD numbers and store the result in BCD. Use the DT directive. See Worksheet question 6.

ACTIVITY 8

Write a subroutine to subtract two 10-digit BCD numbers and store the result in BCD. Use the DT directive. See Worksheet question 7.

1. Demonstrate your understanding of the different formats ASCII, binary and unpacked BCD by filling in the corresponding formats for each of the decimal numbers below in the following table.

Decimal	ASCII (Hex)	Binary	Unpacked BCD
0			
3			
6			
9			
19			
99			

2. For the test data you selected for Activities 1 and 2, write the result of the addition in BCD as the program stored it. Also write it in hex. Verify that the addition was done correctly.

3. For the test data you selected for Activities 3 and 4, write the result of the subtraction in BCD. Also write it in hex. Verify that the subtraction was done correctly.

4. Verify the sum obtained in the addition in Activity 5. Attach a trace of the execution to show your program obtained the same result.

5. Verify the result obtained in the subtraction in Activity 6. Attach a trace of the execution to show your program obtained the same result.

6. Verify the sum obtained in the addition in Activity 7. Attach a trace of the execution to show your program obtained the same result.

7. Verify the result obtained in the subtraction in Activity 8. Attach a trace of the execution to show your program obtained the same result.

LAB 13

USING INT 10H AND INT 21H

OBJECTIVE

» To use INT 10H and INT 21H to perform screen I/O

REFERENCES

Mazidi, Volumes I & II: Chapter 4

MATERIALS

80x86 IBM (or compatible) computer
MASM (or compatible) assembler
DEBUG program

ACTIVITY 1

Write a program which contains Assembly language instructions to clear the screen.

(a) Add to the above program instructions to set the cursor at row=3, column=4, display the prompt "What is your name?" and receive the user's response. Then the program should display the response at row=5, column=6.

(b) Add instructions to set the cursor at row=7, column=4, display the prompt "What is your ID number?" and receive the user's response, then display it at row=9, column=6.

(c) Add instructions to set the cursor at row=11, column=4, display the prompt "What is your address?" and receive the user's response, then display it at row=13, column=6.

(d) Add instructions to set the cursor at row=15, column=4, display the prompt "What is your birthday?", receive the user's response, then display it at row=17, column=6.

(e) Add Assembly language instructions to set the cursor at row=19, column=4, display the prompt "Where were you born?", receive the user's response, then display it at row=21, column=6. See Worksheet questions 1 and 2.

Save the program you have written, it will be used in Lab 14.

1. For Activity 1, dump the data segment before and after running the program, make a print-out of it, then circle the final result in the data segment.

2. When using the string input function 0AH of INT 21H, is a carriage return stored as part of the input string? Underscore the input string for the user's name, then place a box around the number of bytes requested, and circle the number of bytes actually input. Is a carriage return included in the number of bytes actually input?

3. What is the maximum number of characters that can be input (excluding carriage return) for the following buffer?

   DATA_BUF   DB   12,?,13 DUP (?)

LAB 14

MACROS

OBJECTIVE

» To use macros to write more efficient Assembly language programs

REFERENCES

Mazidi, Volumes I & II: Chapter 5

MATERIALS

80x86 IBM (or compatible) computer
MASM (or compatible) assembler

ACTIVITY 1

Rewrite the Program written for Lab 13 using a macro for displaying prompts, another macro for getting a response from the user, and other for setting the cursor position. Use .LALL for part (a), .XALL for parts (b), (c), (d), and .SALL for part (e). Run the program to make sure it works. See Worksheet question 1.

1. Attach the list file generated when you assembled the program. Highlight and comment on the effect of three listing directives on your listing file. Specifically, what effect did LALL, SALL, and XALL each have in terms of comments preceded by a single colon, comments preceded by a double colon, comments on the same line as code, and expansion of macro instructions in the list file?

LAB 15

GRAPHICS MACROS

OBJECTIVE

» To use the concept of macros to make graphics programming easier

REFERENCES

Mazidi, Volumes I & II: Chapter 4, Section 4.1, Chapter 5

MATERIALS

80x86 IBM (or compatible) computer
MASM (or compatible) assembler
DEBUG program

ACTIVITY 1

Write and test a macro to draw a horizontal line. The macro should have 3 arguments: one for the starting column, one for the ending column, one for the row. See Example 4-5 of the textbook for a reminder on how to draw a line. See Worksheet question 1.

ACTIVITY 2

Write and test a macro to draw a vertical line. The macro should have 3 arguments: one for the starting row, one for the ending row, one for the column.

ACTIVITY 3

Write a program to draw a tic-tac-toe symbol on the screen using the above two macros. Use whatever graphics mode you like (medium or high resolution, 4 or more colors). Remember that to reset your monitor to text mode enter "mode co80" at the DOS prompt (letters co followed by numbers 80). While testing the program you might want to use INT 21H Function 07 (get one character, no echo) at the end of the program, so that you can see the tic-tac-toe symbol before the DOS prompt reappears. See Worksheet questions 2 and 3.

1. Draw a flowchart representing the algorithm you used in the macro you wrote for Activity 1.

2. Did you declare program labels LOCAL in your macros? If not, what happened when you called them more than once from the program you wrote for Activity 3? Attach a list file showing the expansion of all macros. Circle any local labels in the macro expansions.

3. How did you determine the beginning and ending points for your lines? Did you code these points into the code segment or did you use data labels? What advantage is there in setting up the line row and column information in the data segment as opposed to the code segment?

LAB 16

SIGNED NUMBER ARITHMETIC

OBJECTIVES

» To code Assembly language instructions to perform signed addition and subtraction
» To use the IMUL instruction to perform signed multiplication
» To use the IDIV instruction to perform signed division

REFERENCES

Mazidi, Volumes I & II: Chapter 6, Section 6.1

MATERIALS

80x86 IBM (or compatible) computer
MASM (or compatible) assembler
DEBUG program

ACTIVITY 1

Fill in the table for Worksheet question 1 by executing the indicated instructions in DEBUG and observing the results.

ACTIVITY 2

Write a subroutine that finds the lowest, the highest, and the average temperature for a given week. Make sure you have some negative temperature in your data. See Worksheet question 2.

ACTIVITY 3

Assume a given voltage reading in millivolts has the range of -12000 to + 12000. Write a subroutine to find the lowest, the highest, and the average reading. Use the data below. See Worksheet question 3.

READ_MIL_VOLT   DW   -1568,8756,900,-11986,2938,543,-3103,33,-99

1. Fill in the following table by executing the indicated instructions in DEBUG and observing the results. For the Result column, indicate the register(s) which hold the result as well as what the result is.

Instruction	Destination	Source	Result	CF	SF	AF	OF	ZF
SUB AL,BL	AL = +46H	BL =+64H						
ADD CH,BL	CH = -5	BL =+1						
SUB DH,BL	DH = -1	BL = +1						
DIV BL	AL = 0	BL=+1						
MUL BX	AX =+ 3934H	BX =-7000H						
DIV BX	AX = +1234H	BX = +1234H						

2. Based on your observations in Activity 2, answer the following questions about the instruction "CMP AL,BL".

   (a) What will be the flag settings if AL > BL?

   (b) What will be the flag settings if AL < BL?

   (c) What will be the flag settings if AL = BL?

3. For Activity 3, attach a dump of the data segment before and after the program run. Circle the storage of the highest, the lowest, and the average readings. Write next to each its decimal equivalent.

LAB 17

STRING OPERATIONS

OBJECTIVE

» To use the STOS and LODS instructions in a data transfer program

REFERENCES

Mazidi, Volumes I & II: Chapter 6, Section 6.2

MATERIALS

80x86 IBM (or compatible) computer
MASM (or compatible) assembler

ACTIVITY 1

Write a program that clears the screen.

Add instructions to display a prompt such as "Type in the message" and wait for the user to respond.

Add instruction to get the user's typed-in string and save it.

Add instructions to display another prompt such as "Would you like to convert from the upper case to lower case Y/N?".

If the user types in "Y" or "y" then the message typed in by the user is converted to lower case and displayed; otherwise, the program ends and returns to DOS. The conversion must be done using the LODS and STOS instructions. This can done by modifying Program 3-4 in Chapter 3 of the textbook.

1. For the instruction STOS, state the implied source and destination operands. For the instruction LODS, state the implied source and destination operands. Did you use a REP prefix in your program? What condition(s) will cause the repeat to stop?

2. Attach a screen print (use Shift-PrtScr) of the screen after the user's input has been converted to upper case and displayed.

LAB 18

MORE STRING INSTRUCTIONS

OBJECTIVE

» To use the SCANS and CMPS instructions in string processing programs

REFERENCES

Mazidi, Volumes I & II: Chapter 6, Section 6.2

MATERIALS

80x86 IBM (or compatible) computer
MASM (or compatible) assembler
DEBUG program

ACTIVITY 1

Write and run a program that displays a set of characters (letters and numbers) and then waits for the user to type in the identical sequence of characters. If they match it will go back to DOS, otherwise it should prompt the user to try it again. See Worksheet question 2.

ACTIVITY 2

Write and run a program that displays a message such as "I Always wanted to gO to the orient", then asks the question "Would you like to scan and replace a letter (Y/N)?" If the answer is N, then it goes back to DOS. If the answer is Y, then it should display the prompt "Type in the single letter you want to scan first, followed by the letter you want to replace". The program should replace the letter and display the corrected message, then return to the Y/N prompt until the user types in N.

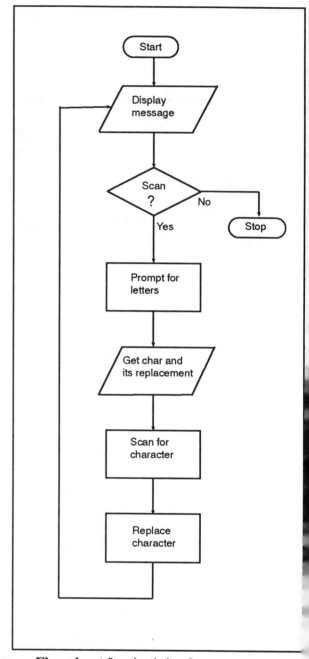

Flowchart for Activity 2

1. For the instruction SCAS, state the implied source and destination operands. For the instruction CMPS, state the implied source and destination operands.

2. For Activity 1, dump the data segment before and after the program run. Attach a screen print of the screen showing a user response which is not identical, and one which is.

3. For Activity 2, dump the data segment before and after the program run. Attach a screen print of the screen showing at least one character replaced.

LAB 19

USING A LOOK-UP TABLE

OBJECTIVE

» To use the XLAT instruction in a table processing program

REFERENCES

Mazidi, Volumes I & II: Chapter 6, Section 6.2

MATERIALS

80x86 IBM (or compatible) computer
MASM (or compatible) assembler

ACTIVITY

A given PC is connected to a seven segment LED. Using a look-up table, write a program that will convert any key representing a hex number (0, 1, 2, ..., 9, A, B, ..., F) to its 7-segment displayable format. For example, to display "2" on a 7-segment LED 06 must be sent to the LED circuitry as shown next.

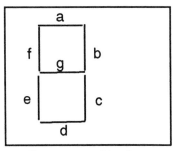

7-segment display

g	f	e	d	c	b	a
1	0	1	1	0	1	1

= 5B HEX

ACTIVITY 2

Your program should take in a single input and examine it to make sure it is in the range of hex digits of 0-F. If it is, translate it to the 7-segment equivalent and store it; otherwise, it should display a message such as "Your input is not a hex number". See Worksheet question 1.

1. Test the program on the following data.

   A, 0, 1, 5, 10, H

   Trace through the program for the above data, showing the result of each XLAT on the data. Attach the trace.

LAB 20

ADDITION AND SUBTRACTION MODULES

OBJECTIVE

» To write external subroutines which can be linked with other object modules

REFERENCES

Mazidi, Volumes I & II: Chapter 7, Sections 7.1 through 7.2

MATERIALS

80x86 IBM (or compatible) computer
MASM (or compatible) assembler
Turbo C

ACTIVITY 1

Write, assemble, run, and analyze a program that calls modules which perform the tasks outlined below.

ACTIVITY 2

Write a module to find the total sum of many individual bytes and store the word-sized result. The choice of the number of bytes and the data is up to you, but the total sum must be more than one byte.

ACTIVITY 3

Write a module to find the sum of two multibyte numbers. Each number should be of 8-byte size. The addition should be done one byte at a time.

ACTIVITY 4

Write a module to find the total sum of many individual words and store the result which will be a double word. The choice of the number of words and the data is up to you, but the total sum must be more than one word.

ACTIVITY 5

Write a module to find the sum of two multibyte numbers. Each number should be of 8-byte size. The addition should be done one word at a time.

ACTIVITY 6

Write a module to subtract one multibyte (8-byte) number from another and save the result.

ACTIVITY 7

Write a module to add two BCD numbers and store the result in BCD. Use the DT directive.

ACTIVITY 8

Write a module to subtract two BCD numbers and store the result in BCD. Use the DT directive.

Save this program because it will be used in Lab 21.

ACTIVITY 9

Write the main (entry) section of this lab in C. Use some of the Assembly modules from Activity 8 to create an executable program to be run from DOS. In the main portion (written in C), the user is prompted for the data. Then the data is passed to the modules (written in Assembly). The results are returned to the main module to be displayed.

Note: The main portion must be saved with the extension C, not the usual CPP (Ex: LAB20.C).

1. Write down the link command you used to link the modules together. How many modules did you use in how many separate files? Were your modules FAR or NEAR? State the differences in NEAR and FAR modules in terms of how the CALL instruction is affected. Unassemble the CALL instruction for the NEAR call and print it out. Change the modules to NEAR if you used FAR, or to FAR if you used NEAR, then reassemble and relink the program. Unassemble the CALL instruction for the FAR call and print it out. Did the NEAR and FAR CALL instructions have the same machine code? What differences in the operands did you notice?

LAB 21

ANALYZING THE LINK MAP

OBJECTIVE

» To use the ".map" file generated by the linker to learn more about segments and memory management

REFERENCES

Mazidi, Volumes I & II: Chapter 7, Section 7.1 through 7.2

MATERIALS

80x86 IBM (or compatible) computer
DEBUG program
MASM (or compatible) assembler

THE MAP FILE

The "/m" option with the LINK command generates an expanded map file.
C>LINK /M PROG1.OBJ + MOD1.OBJ + MOD2.OBJ

This will generate a map file (default name PROG1.MAP) which, in addition to the segment information normally given in a map file, will also contain a list of all symbols in the program. The symbols will be listed alphabetically as well as by offset address.

ACTIVITY 1

Analyze the link map generated in the solution to Lab 20. All segments should have PARA boundary and be defined as PUBLIC. Each code segment module should have a different name.

ACTIVITY 2

Analyze the link map generated in the solution to Lab 20 with the first 4 module segments having PARA boundary and the rest WORD boundary and defined as PUBLIC. Each code segment should have its own unique name.

ACTIVITY 3

Analyze the link map generated in the solution to Lab 20 with all segments having PARA boundary and defined as PUBLIC. Each code segment module should have the same name. Use the "/m" option to generate a map file listing all symbols in the program. Also generate a list file.

1. Contrast and compare the link maps for each of Activities 1, 2, and 3 in terms of beginning and ending segment addresses and segment size.

2. Comment on the information given in the expanded map file when the program was linked with the "/m" option. Compare the offset addresses in the map file with the offsets of the data segment in the list file.

LAB 22

ASCII ADDITION/SUBTRACTION MODULES

OBJECTIVE

» To write modules to perform ASCII addition and subtraction

REFERENCES

Mazidi, Volumes I & II: Chapter 7, Section 7.2

MATERIALS

80x86 IBM (or compatible) computer
DEBUG program
MASM (or compatible) assembler

ACTIVITY 1

Write a program that displays a prompt such as "Enter the first number", then gets the number. Then the program should prompt for the second number. The number can be any length up to 16 digits. The numbers do not have to be the same length.

ACTIVITY 2

Write a module that adds the two numbers. The main module will display the sum with a line such as "The sum is ". The addition must be performed in ASCII.

ACTIVITY 3

Write a module that subtracts the two numbers. The main module will display the result with a line such as "The difference is ". The subtraction must be performed in ASCII.

1. Attach a dump of the data segment before and after the program run. Verify that the modules added or subtracted the data correctly. Use Appendix B.2 in the textbook to calculate the clock count for one iteration of the loop. Then calculate the clock count for a program which would convert the numbers to hex (using the conversion subroutines shown in Programs 7-6 and 7-7 of the textbook), add them, then convert them back to ASCII.

LAB 23

CREATING AND USING LIBRARY FILES

OBJECTIVES

» To use the LIB command to create a library of modules
» To link an object file with modules contained in a library

REFERENCES

Mazidi, Volumes I & II: Chapter 7, Section 7.2

MATERIALS

80x86 IBM (or compatible) computer
MASM (or compatible) assembler

CREATING LIBRARY FILES

A library file is a collection of several object files. This makes it easier to call the procedures from different programs. For example, assume that you wish to combine various modules which perform arithmetic into one library so that they could be called from various programs. First the programs must be assembled with MASM to product the ".obj" files. Then the object files can be combined into a library with the LIB command as follows:

C>LIB ARITH +ADD + SUB;

This command will created a library named ARITH.LIB which contains the two object modules ADD.OBJ and SUB.OBJ. Other modules can be added later as follows:

C>LIB ARITH +MUL +DIV,ARITH.LST

The above command adds object modules MUL.OBJ and DIV.OBJ to ARITH.LIB and also produces a list file named ARITH.LST which will list information about the modules included in the library.

The following shows the commands to delete a module. This will remove MUL.OBJ from the library ARITH.

C>LIB ARITH -MUL;

The following shows how to replace a module. The module ADD in library ARITH will be replaced with ADD.OBJ in the current directory. This command would be used if ADD.ASM had been modified and reassembled, and you wish to replace the old version in the library with the new version of ADD.OBJ.

C>LIB ARITH -+ADD;

The following shows how to copy a module from the library to the current directory. This command will extract module SUB.OBJ from library ARITH, and place it in the current directory. The module will not be deleted from the library.

C>LIB ARITH *SUB;

Finally, the following shows how to extract a module from the library and delete it from the library. The following would delete module DIV from library ARITH and place DIV.OBJ in the current directory.

C>LIB ARITH -*DIV;

Linking with library files

The following command links PROG1.OBJ with modules in ARITH.LIB to produce PROG1.EXE. It is necessary for ARITH.LIB to be in the same directory as PROG1.OBJ, and the modules in PROG1.ASM must have been declared as EXTERNAL.

C>LINK PROG1,,,ARITH;

ACTIVITY 1

Combine the ASCII-to-hex and the hex-to-ASCII modules (Programs 7-6 and 7-7 of the textbook) into one library called CONVERT.LIB. Make a list file of the library. Then link the library with Program 7-8 on page 186. Run the program to verify that it runs correctly. See Worksheet question 1.

1. Attach a copy of the LIB list file generated when the library was created. List the information given in the LIB list file and state what it means. List below the command you used to generate the LIB file. List also the command you used to LINK the modules together.

LAB 24

32-BIT PROGRAMMING

OBJECTIVE

» To use the 32-bit registers in Assembly language programs

REFERENCES

Mazidi, Volumes I & II: Chapter 8, Sections 8.1, 8.2

MATERIALS

80x86 IBM (or compatible) computer
MASM (or compatible) assembler
Microsoft's CodeView or Borland's Turbo Debugger

ACTIVITY 1

Write a program to subtract two 8-byte numbers using 32-bit registers. Your program should use a loop (see Program 8-3b in the textbook). Use the following data. See Worksheet question 1.

```
DATA1 DQ    548FB9963CE7H
DATA2 DQ    3FCD4FA23B8DH
```

ACTIVITY 2

Write a program to subtract two 8-byte numbers using 32-bit registers. Your program should not use a loop (see Program 8-4 in the textbook). Use the data given in Activity 2. See Worksheet question 2.

ACTIVITY 3

Write a program similar to Program 8-5 to divide a 32-bit operand by a 16-bit operand. Use the following data. See Worksheet question 3.

```
DATA1 DD    500000
DATA2 DD    50000
```

1. For Activity 1, use Shift-PrtScreen to display the results of the program and the contents of the 32-bit registers.

2. For Activity 2, use Shift-PrtScreen to display the results of the program and the contents of the 32-bit registers.

2. For Activity 3, use Shift-PrtScreen to display the results of the program and the contents of the 32-bit registers.

SECTION 2

Advanced Labs

The 10 labs on the following pages are advanced labs which can be used as special projects, or for supplemental labs to challenge more advanced students. The labs are listed in order below, along with the chapter after which the student should be able to complete the lab.

Advanced Lab	After Chapter
1. Bubble Sort Program	3
2. Fibonacci Sequence Program	4
3. Date/Time Display Program	4
4. Line Draw Program	4
5. Box Draw Program	5
6. Register Dump Module	4
7. Phone Directory Program	6
8. Character Encoding Program	6
9. Hex Calculator Program	7
10. Data Entry Program	7

LAB 1

BUBBLE SORT PROGRAM

A bubble sort algorithm loops through a set of data items, comparing successive items and switching them if they are in the wrong order. This process is repeated until the list is sorted. It is called a bubble sort because the smaller items "bubble up" towards the top of the list. The bubble sort algorithm in pseudocode is given below. If the inner loop swaps any items, it should set a register or a memory location to some value which indicates that a swap occurred. The outer loop then checks this "flag" to see if any items were swapped. If items were swapped, the outer loop sets the "flag" to indicate no items were swapped and the inner loop is executed again. This process continues until the outer loop determines that no items were swapped on the inner loop.

```
REPEAT
        REPEAT
                compare successive items
                swap if in wrong order
        UNTIL (end of list)
UNTIL (no items were swapped on inner loop)
```

The following shows how a bubble sort would work on the first loop through a list of five bytes. In the first loop through the data, 93 is compared to 30, they are swapped and 30 will be placed in the first position and 93 in the second. Then 93 and 29 are compared. Since 93 is greater than 29, they are swapped, with 29 being placed in the second position and 93 in the third. Then 93 is compared to 05 in the fourth position, and they are also swapped. Then 93 is compared to 88 in the fifth position, and they are also swapped.

93	30	29	05	88	;original list
30	93	29	05	88	;first and second items compared
30	29	93	05	88	;second and third items compared
30	29	05	93	88	;third and fourth items compared
30	29	05	88	93	;fourth and fifth items compared
30	29	05	88	93	;after first loop through data

The following shows the original list, followed by the way the list would look after each pass through the data. This list was sorted after 3 passes through the data. The number of passes required to sort a list varies from list to list.

93	30	29	05	88	;original list
30	29	05	88	93	;after first loop through data
29	05	30	88	93	;after second loop through data
05	29	30	88	93	;after third loop through data

Write a program to perform a bubble sort on ten bytes of numeric data. Test the program and stop the trace at each iteration of the outer loop to see how each sort through the list affected the data. The XCHG instruction (which swaps the contents of a memory location and a register) might prove helpful. For information on this instruction, refer to Appendix B in the textbook.

LAB 2

FIBONACCI SEQUENCE PROGRAM

The Fibonacci sequence is a series of numbers that starts with 0 and 1. All subsequent numbers are generated by adding the previous two items. For example:

0, 1, 1, 2, 3, 5, 8, 13, 21, 34, 55, 89

The first two items, 0 and 1, are added to get the third item, 1. Then 1 and 1 are added to get 2, 1 and 2 are added to get 3, 2 and 3 are added to get 5, etc.

Write a program which generates the Fibonacci sequence for 15 numbers. The numbers should be stored in consecutive memory locations. Your program should initialize the first two locations to 0 and 1, respectively, then repeatedly call a subroutine which computes the next number in the series and stores it in memory. After all the numbers have been generated, the series should then be displayed on the screen.

Store 0 and 1 in their ASCII forms, 30H and 31H. Add the ASCII numbers and use the AAA instruction. For information on instruction AAA, refer to Appendix B or Section 3.4 of the textbook. In setting aside storage locations for the series, you must anticipate the number of bytes needed to store the largest item.

LAB 3

DATE/TIME DISPLAY PROGRAM

DOS INT 21H Function 2AH gets the system date. The information is returned in the following registers:

CX   year (1980 through 2099)
DH   month (1 through 12)
DL   day (1 through 31)
AL   day of the week (0 = Sunday, 1 = Monday, ..., 6 = Saturday)

DOS INT 21H Function 2CH gets the system time. The information is returned in the following registers:

CH   hours (o through 23)
CL   minutes (0 through 59)
DH   seconds (0 through 59)
DL   hundredths of seconds (0 through 99)

Write a program that displays the system date and time on the screen. Place the information within a graphics box. A table might be used to get the day of the week based on the value in AL. The table might looks as follows:

```
MONTH_TBL DB    'January  '
          DB    'February '
          ...
          DB    'December '
```

Each entry in the table would be 9 bytes long. Therefore the starting offset address of any month could be calculated by the following formula, offset = (month - 1) × 9. For example, the offset for month 1 would be (1 - 1) × 9 = 0. For month 2, the offset would be (2 - 1) × 9. Once the offset for the beginning of the string is computed, the 9 bytes containing the name of the month could be moved to the display area.

Another challenge in this program is to transform a register value, such as the hex byte CL which holds the minutes, let's say 34H (52 decimal), into the ASCII representation of decimal 52, "35 32". One way would be as shown below. This code copies CL to AL, zeroes AH, divides AL by 0AH. This puts the quotient, 05, in AL and the remainder, 02, in AH.

```
MOV AL,CL          ;copy minutes to AL
MOV AH,0           ;set up for divide
MOV BL,0AH         ;divide by 10 to convert to decimal
DIV BL             ;now AL = 05 and AL = 02 (remainder)
OR AX,3030H        ;make it ASCII
MOV ...,AL         ;move 35H to lower byte of minutes in buffer
MOV ...,AH         ;move 32H to upper byte of minutes in buffer
```

LAB 4

LINE DRAW PROGRAM

Write a simple line draw program. The program should first set the monitor to text mode, then clear the screen. Next it should set the cursor to row = 1, column = 1, and display '*'. Next the user should be able to draw lines of stars by entering any of the four arrow keys. In other words, if the user enters a right arrow, draw a star one position right; if the user enters three down arrows in a row, draw three stars downward, etc. The user can enter letter "Q" to stop the program. Any other keys can be ignored. The four ASCII codes for the arrows are as follows:

left arrow	4BH
right arrow	4DH
up arrow	48H
down arrow	50H

When entering these scan codes, the "get character" function should be called twice (as shown below) because the system first sends a 00 to indicate that an extended scan code is being sent, then it sends the hex code.

```
MOV AH,07    ;get extended code 00
INT 21H
MOV AH,07    ;get char
INT 21H
```

LAB 5

BOX DRAW PROGRAM

Write a program to draw three concentric boxes. Draw the outer box first, from the top left corner, clockwise, then draw the middle box counter-clockwise, and finally the inner box clockwise. Insert a SLOW module (shown below) to make the system draw the boxes in slow motion. Choose 3 different colors for your boxes. The following shows a structure that could be used to define the starting and ending coordinates of a line. It is followed by four definitions of lines with their initial data. These four lines will comprise a box when drawn. Refer to Appendix C in the textbook for more information about structures.

```
LINE        STRUC              ;begin line structure
ROW_STRT    DW    ?
COL_STRT    DW    ?
ROW_END     DW    ?
COL_END     DW    ?
LINE        ENDS         ;end line structure

LINE1       LINE <10,10,10,309> ;starts at 10,10 ends at 10,309
LINE2       LINE <10,309,189,309>        ;starts at 10,309 ends at 189,309
LINE3       LINE <189,10,189,309>
LINE4       LINE <10,10,189,10>
```

Write 4 macros: one to draw a horizontal line from left to right, one to draw a horizontal line from right to left, one to draw a vertical line from top to bottom, one to draw a vertical line from bottom to top. The following is a macro that would use the data definitions above to draw a horizontal line from left to right.

```
HORIZ_L_R   MACRO LINE_H
            LOCAL DRAW_L_R
;this macro draws a horizontal line from left to right
            MOV   BX,LINE_H.COL_END
            MOV   AH,0CH                 ;draw pixel option
            MOV   AL,10B                 ;pixel value
            MOV   DX,LINE_H.ROW_STRT     ;DX = row
            MOV   CX,LINE_H.COL_STRT     ;CX = column
DRAW_L_R:   INT   10H                    ;draw a pixel
```

```
          SLOW                           ;slow down
          INC     CX                      ;move 1 column to the right
          CMP     CX,BX                   ;is it the last column?
          JB      DRAW_L_R                ;if not, continue drawing
          ENDM
```

The above macro would be called with LINE1 as follows:

```
          HORIZ_L_R   LINE1
```
There are many ways to write the SLOW macro; here is one way:

```
SLOW          MACRO
              LOCAL       SLOW_LOOP
;this macro inserts a delay into program by counting down from FFFF to 0
;this is slow enough for a 50 MHz CPU, it may be too slow for other CPUs
              MOV   SI,0FFFFH
SLOW_LOOP:SUB   SI,1
              JNZ   SLOW_LOOP
              ENDM
```

LAB 6

REGISTER DUMP MODULE

Write a module to print a register dump to the screen whenever it is called. The dump should be similar to the DEBUG register dump, as shown below.

Test the routine by calling it several times in one of your programs. Note that if you want to display the contents of CS and IP within a module, these values will have been changed when the module was called; therefore, the original values should be retrieved from the stack. Refer to Section 7.3 in the textbook to refresh your memory on how CS and IP are stored on the stack prior to subroutine calls.

There are many ways to approach this problem, but one way is to build a shell of each display line (shown above) in the data segment, plug in the register values when the module is called, and print it with the display function INT 21H Function 09. Part of the shell for the first line might look as follows:

```
LINE_1      DB      'AX='
REG_AX      DB      4 DUP (' 0')
            DB      ' BX='
REG_BX      DB      4 DUP (' 0')
            ...
            ...
REG_DI      DB      4 DUP (' 0')
            DB      '$'                   ;terminates display
```

The challenge in this program is to transform a register value, which is a hex word such as "12BC", into its ASCII representation, in this case "31 32 42 43". One way is to first save the register (let's assume that it is AX) in a memory location or on the stack. Then mask off all but the first hex digit of AX, rotate that digit to the least significant digit position, and use it as an offset into an XLAT table to get its ASCII equivalent. That ASCII code then could be stored at REG_AX. Next, restore the original value of AX, mask off all but the second digit, rotate that digit to the least significant digit, gets its hex equivalent from the table, and store it at REG_AX+1. This must be done for the remaining two digits as well to translate the hex word into 4 ASCII bytes. Of course, any registers that you would use for the rotate, XLAT, or other instructions would have to be saved on the stack to preserve the values they had when the routine was called.

LAB 7

PHONE DIRECTORY PROGRAM

AX=15DE BX=0000 CX=005D DX=0000 SP=0000 BP=0000 SI=0000 DI=0000
DS=15DE ES=15CC SS=15DC CS=15DF IP=0005   NV UP DI PL NZ NA PO NC

Write a program to search a table that contains the first names and phone numbers of up to 20 of your friends. The table might look something like this:

```
FRIENDS     DB      'JANIE   555-0302$'
            DB      'GEORGE  555-3212$'
            DB      'LUPITA  555-2034$'
            DB      'JOHN    555-2304$'
            DB      'CHEN    555-2213$'
            DB      'MONA    555-2344$'
            . . .
            . . .
            DB      'ALI     555-9009$'
```

The user should be prompted for a name which can be up to 6 characters long. The program will search the table for the name and print the phone number if found, or print an error message if it is not found. The search could be performed as follows. The offset address of FRIENDS is placed in BX. The input item is compared with the item at offset BX by use of the CMPSB instruction and a REP prefix. If the two strings are equal, that offset will be used with the display string function (INT 21H function 09) to print the phone number to the screen. If the two items do not match, 16 is added to BX to make it point to the next item in the table. This process continues, as the program loops through the table, until a match is found or until the end of the table is reached.

LAB 8

CHARACTER ENCODING PROGRAM

An encoding system is one which translates a string of characters into an encoded string which cannot be read by unauthorized users. Decoding is the opposite process of taking an encoded string and translating it into its readable form. For example, the following tables could be used to form a coding pattern for ASCII uppercase letters.

```
ENCODE_TABLE      DB      'DIBOUVSFAZKMCQGRNTPJWEXLYH'
DECODE_TABLE      DB      'ABCDEFGHIJKLMNOPQRSTUVWXYZ'
```

Using the above encoding system, the string 'GOOD MORNING' would become 'SGGO CGTQAQS'. For example, to encode the first letter of the original string, search DECODE_TABLE for the letter 'G' to find its position. The letter 'G' is the 6th element (remember to start counting at 0). Therefore, 6 could be placed in AL prior to the execution of an XLAT into the encoding table. The 6th element of ENCODE_TABLE is 'S'. Thus 'G' is encoded into 'S'.

Write an encoding routine for ASCII digits 0 - 9, uppercase letters and lowercase letters. Write a corresponding decoding routine. Test the program by taking input from the keyboard, encoding it, writing it to a buffer, decoding it, and echoing back both the original message and the encoded message.

LAB 9

SIMPLE CALCULATOR PROGRAM

Write a program which prompts the user to enter two numbers up to 65,535 in value, and an operation to be performed on the numbers (+, -, *, /). The program will then display numbers and the result, as shown below.

```
ENTER FIRST NUMBER:43981
ENTER SECOND NUMBER:4660
ENTER OPERATION:+


     43981
      4660

SUM:  48641
```

LAB 10

DATA ENTRY PROGRAM

Write a program which accepts up to 10 numbers from the keyboard. Negative numbers will be preceded by a minus sign; positive numbers may or may not be preceded by a plus sign. The numbers can range in value from -32,768 to +32,767. Find the average, the highest, and the lowest items. After the items are input, clear the screen, display the items in a right-justified column as shown below, along with the average, highest, and lowest items.

```
       +5
    -32700
     -4623
    +12965
      +369
      -999
     +5402
    +19805
     +6245
       -15

The average is 645.
The highest is 19805.
The lowest is -32700.
```

SECTION 3

System Programming

This section uses Assembly language to explore the architecture and operating system of the IBM PC. Assembly language programs are written to determine which CPU is installed, the PC RAM size, to examine the BIOS data area and interrupt assignments. Other labs in this section experiment with playing music on the PC, keyboard input, video programming, and file processing on floppy and hard disks. Further topics include 32-bit programming, memory management, and TSR programming.

LAB 1

FINDING WHICH 80x86 IS INSTALLED

Objective

» To write and test a program that identifies the 80x86 CPU on the motherboard

References

Mazidi, Volumes I & II: Chapters 2, 3, 6, 23
Intel, Pentium Processor User's Manual Vol. 3, Chapter 5
Leinecker, Richard, "Processor Detection Schemes," Dr. Dobb's Journal #201, Jan. 1993.

Materials

80x86 IBM (or compatible) computer
MASM (or compatible) assembler
MicroSoft CodeView (or Borland's Turbo Debugger)

Activity 1

Write a program that identifies the x86 processor installed in any PC. Upon running your program it should display a message such as "The microprocessor in this PC is 80286".

Note: To assemble the code below you need MASM 6.11 (or TASM 4.0) which support the Pentium instructions such as CPUID (the directive .586 is for that purpose). If you are using MASM 5.x then remove both the .586 directive and CPUID instruction and replace them with opcode for CPUID which is (0FA2H) in the following manner

```
DW      0FA2H          ;opcode for CPUID instruction
```

Every two or three years, a new generation of x86 is introduced. However, if we wish not to update our assembler to support the new instructions, we can use a macro to generate the opcode for the new instructions. For example, earlier versions of MASM do not support new instructions such as CPUID of the Pentium chip. In that case, we can use the following macro instead.

```
CPUID MACRO
        DB      0FH     ;opcode for CPUID instruction
        DB      0A2H
        ENDM
```

Now whenever "CPUID" is encountered in the assembly language program, the opcode for CPUID will be inserted in the assembled program.

See also Jeff Prosise's article in *PC Magazine*, Jan. 24, 1995 to see the use of CPUID in identifying the 486DX4. We recommend reading his article regularly.

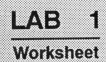
1. Go to the DOS directory and execute the MSD program. It gives you the processor installed on the PC. Does it confirm the result of your program?

2. Which microprocessor supports the CPUID instruction?

3. Which bits of the flag register indicate if the CPUID instruction is supported?

4. State the difference between the PUSHF and PUSHFD instructions.

5. State the difference between the POP and POPFD instructions.

LAB 2

EXPLORING THE BIOS DATA AREA

Objectives

» To examine the contents of the BIOS data area (BDA)
» To write a program to display the installed parallel and serial ports of the PC and their I/O base addresses
» To write a program to produce a list of installed equipment on a PC

References

Mazidi, Volumes I & II: Chapters 17, 28, Appendices E.2, H

Introduction

The BIOS data area of the PC is located at physical addresses 400H - 4FFH (logical addresses 40:00 - 40:FF). When the PC is turned on, the power-on-self-test (POST) subroutines, which are stored in the PC's BIOS ROM, test all peripherals and memory of the system and write some critical information into the BIOS data area. DOS uses the information supplied by the BIOS data area to navigate the system hardware.

A complete listing of the BIOS data area is provided in Appendix H of the textbook. In this lab we explore the contents of locations 400H - 411H. By examining these locations we see that the base port address for each of 4 serial ports is stored in memory locations 40:0000 - 40:0007, where each two locations hold the port address of a single COM port (e.g., 40:0000 and 40:0001 hold the base port address for the serial communication port number 1). See Example 17-3 in the textbook. Locations 40:0008 - 40:000F hold the base port address of printers attached to the PC. Again, each two locations hold the port address of a single LPT port (e.g., 40:0008 and 40:0009 hold the base port address for LPT1). Locations 40:0010 and 40:0011 hold a 16-bit word indicating the system equipment installed on the PC (This information is also returned in AX by running interrupt instruction INT 11H. See Appendix E.2).

Activity 1

Use DEBUG's DUMP command to dump memory location F000:FFF5 to see the date of your BIOS as follows:

```
C:\dos\debug
-d f000:fff5
```

See worksheet question 1.

Activity 2

Use DEBUG's DUMP command to dump BIOS data area 40:0000 -
40:0011 and examine the port addresses assigned to the serial (COM) and
parallel (LPT) devices. Also examine the equipment word. See worksheet
questions 2 - 4.

Activity 3

Part (a)

Write and run a program (in C or Assembly) to display (1) the base
I/O address for each of the installed COM ports, and (2) the base I/O address
for each of the installed LPT ports.

Note: If the base address is 0 it indicates that the port is not installed.
Your program should display a message such as "COM port 1 is installed and
the base port address is ...". or "COM port 2 is not installed".

Part (b)

Write and run a program (in C or Assembly) to get the equipment
word (either from the BIOS data area or the INT 11H instruction) and display
the information indicated by each bit (e.g., "Math Coprocessor installed").

1. In Activity 1, what is the date of BIOS creation for your computer?

2. Fill in following tables for Activity 2.

   Table 1:

Parallel Port	Installed (yes/no)	Port address (in hex)
LPT1		
LPT2		
LPT3		
LPT4		

   Table 2:

Serial Port	Installed (yes/no)	Port address (in hex)
COM1		
COM2		
COM3		
COM4		

   The equipment word is _____ (hex) _____ (binary).

   Interpret the above equipment word and fill in the following:

_____	Number of printer adapters
_____	Number of serial adapters
_____	Number of diskette drives
_____	Initial video mode
_____	Mouse installed (yes/no)
_____	Math coprocessor installed (yes/no)
_____	Disk drive installed (yes/no)

3. In a DOS directory, run command MSD to see if it confirms the results in Questions 1 and 2.

4. Using DEBUG, assemble and execute INT 11H as follows. *Reminder*: In DEBUG all numbers are assumed to be in hex.

   ```
   INT 11
   INT 3
   ```

   Do you get the same value for the equipment word as in Question 2?

5. In Activity 3, does your program confirm the result indicated by Questions 1 and 2?

LAB 3

FINDING THE RAM SIZE IN A PC

Objectives

» To examine the BIOS data area to find the size of base RAM (conventional memory)
» To write a program to find the amount of base RAM
» To write a program to find the amount of RAM beyond 1M (extended memory)

References

Mazidi, Volumes I & II: Chapters 11, and 28, Appendix H

Introduction

In the PC there are several ways to find out the size of conventional memory (the 640K byte memory):

1. In DOS you can use the MEM command.
2. In DOS you can also use the MSD command.
3. In software programming there are two options:
   (a) Using INT 12H. Upon returning from INT 12H, the AX register contains the RAM size in hexadecimal format.
   (b) Reading the contents of the BIOS data area 40:0013H and 40:0014, as shown next.

```
PUSH DS              ;save DS
MOV  AX,40H          ;set AX=40H the BIOS data area seg
MOV  DS,AX           ;DS=40 BIOS data area seg
MOV  AX,[0013H]      ;get RAM size from 0040:0013&14
POP  DS              ;restore DS
```

Notice that the memory size is contained in a 16-bit word, where location 40:0013 holds the low byte and the high byte is held by 40:0014H. Again, the size is in hex format.

Activity 1

Part (a)

Using DEBUG, dump BIOS data area locations 413H and 414H to get the size of the base RAM memory.

Part (b)

In DEBUG, assemble and execute instruction INT 12H as follows to get the size of base memory.

```
INT 12
INT 3
```

See Question 1 in the worksheet.

Activity 2

Write an assembly language program to display the size of the conventional memory of any 80x86 IBM PC and compatible. Since the memory size is provided in hex format it must be converted from hex to ASCII and then displayed on the screen. For the hex-to-ASCII conversion routine, see Chapter 7 of the textbook.

Activity 3

Example 28-3 in Chapter 28 shows the C language version of Activity 2. Using the C language compiler of your choice, compile and run Example 28-3. See Questions 2 and 3.

Activity 4

Write an Assembly language program to display the size of memory beyond 1M. Notice that while the conventional memory size is provided by the BIOS data area, memory beyond one megabyte (the extended memory) is stored in CMOS RAM. CMOS RAM locations 30H and 31H hold the size of the memory beyond 1M. Example 28-13 in the textbook shows how to access this memory.

Again, since the memory size is provided in hex format it must be converted from hex to ASCII and then displayed on the screen. For the hex-to-ASCII conversion routine, see Chapter 7 of the textbook.

Activity 5

Examples 28-13 and 28-14 of the textbook are C language versions of Activity 4. Using the C language compiler of your choice, compile and run Example 28-13 (or 28-14). See Question 4.

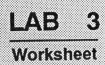
Name: _____
Date: _____
Class: _____

1. In Activity 1 Parts a and b, give the size of base RAM (Kbyte).

   <u>In Hex</u>   <u>In Decimal</u>

   <u>RAM Size</u>
   Activity 1a
   Activity 1b

   Use the MSD utility in DOS to confirm the above result.

2. Are the results of your programs in Activities 2 and 3 confirmed by the result of Question 1?

3. Give the size of the EXE file for both Activities 2 and 3. *Hint:* Use the DIR command.

   Activity 2
       Assembly language EXE file size   = _____ bytes
       C language EXE file size   = _____ bytes

   Activity 3
       Assembly language EXE file size   = _____ bytes
       C language EXE file size   = _____ bytes

4. Repeat Question 3 for Activities 4 and 5.

   Activity 4
       Assembly language EXE file size   = _____ bytes
       C language EXE file size   = _____ bytes

   Activity 5
       Assembly language EXE file size   = _____ bytes
       C language EXE file size   = _____ bytes

   Write your conclusion about the size and efficiency of Assembly versus C language.

LAB 4

PLAYING MUSIC ON A PC

Objective

» To write a program to play music using the PC's 8253/54 timer/speaker

References

Mazidi, Volumes I & II : Chapter 13

Materials

80x86 IBM PC (or compatible) computer
MASM (or compatible) assembler

Activity 1

Write and run an Assembly language program to play the musical notes of the Happy Birthday song. See Chapter 13 of the textbook for the notes. Instead of Happy Birthday, you may use notes from your own favorite song. Use the piano note frequencies in Figure 13-5 of the textbook for your chosen song.

Activity 2

When programming the 8253/54 to play music, counter 2 must be loaded with the value by which 1.19131 MHz needs to be divided. This is shown in Examples 13-5 and 13-6 of the textbook. However, using the method in these two examples requires the programmer to calculate the value for each frequency (by dividing the 1.1931 MHz by the desired frequency). So why not let the 80x86 do the division for us? All we have to do is to make a table of frequencies for the notes and their durations. The program will do the rest. For example the table will look as follows:

```
;from the  data segment
;table of frequencies and durations for each note of 'Happy Birthday'
MUSICDATA  DW    262,1  ;C4 note frequency,duration of 1
           DW    262,1  ;C4 note frequency,duration of 1
           DW    294,2  ;D4 note frequency,duration of 2
           DW    262,1  ;C4 note frequency,duration of 2
           DW    349,2  ;F4 note frequency,duration of 2
           DW    330,4  ;E4 note frequency,duration of 4
;and so on
```

In the code segment, the divisor value to be loaded into counter2 can be calculated by the CPU as follows.

```
;assume SI=offset of MUSICDATA
;assume DX:AX=1,191, 3MHz (numerator to calculate count2 divisor)
           MOV  BX,[SI]       ;get the note frequency
           MOV  CX,[SI+2],     ;get the duration
           DIV  BX            ;calculate count2 divisor value
           CALL  PLAY
```

Repeat Activity 1 using the above method.

Activity 3

Perform Activity 1 by writing a program in C language to prompt the user for the following information: (a) number of notes to be played, and (b) each note and its duration.

After the user finishes entering the data, the following message will appear on the screen: "Type any key to play the song".

Do not use the C function SOUND, since the quality is not as good as programming the 8253/54 directly. For C language programming of the 8253/54, see Example 28-9 in the textbook.

1. Discuss the following methods of time delay generation.
   (a) using a clock count for each instruction
   (b) using a fixed delay by monitoring the hardware bit toggling of I/O port 61H bit 4

2. Show the 50 ms time delay generation using both methods discussed in Activity 1. Assume that the system is a 486 when using the clock count method.

LAB 5

EXAMINING THE PC INTERRUPT ASSIGNMENT AND THE INTERRUPT VECTOR TABLE

Objectives

» To explore the hardware interrupt assignment of the PC using the Microsoft Diagnostics (MSD) utility

» To write an Assembly language program to examine the 80x86's interrupt vector table to see which ones are unused

» To write C program to show the categories of all 256 interrupts

References

Mazidi, Volumes I & II: Chapters 14 and 28

Materials

DOS 5 or higher
MASM or compatible assembler
Turbo C (or compatible) compiler

Activity 1

Using the MSD utility (this comes with DOS), examine the assignment of hardware interrupts in a PC. See Questions 1 - 3.

Activity 2

Of the 256 interrupts of the 80x86, some are used by BIOS, some are used by DOS, and the rest are unused. In examining the interrupt vector table, if the entry for the interrupt is zero, that interrupt is not used by DOS or BIOS and therefore it is unassigned. Write an Assembly language program to display the list of all the unused interrupts in the following format:

UNUSED INTERRUPTS (BY INT#)
..
..

One approach to this program is code it so that it can be run on any 80x86 PC. The contents of the interrupt vector table for each interrupt entry are moved into BX:AX and tested to see if they all zeros.

```
;assuming ES=0, SI=0 (pointers to interrupt vector table)
BACK: MOV   AX,ES:[SI]      ;get the offset value
      ADD   SI,2            ;point to segment value
      MOV   BX,ES:[SI]      ;get the segment value
      ADD   SI,2            ;point to offset of next entry
      COMP  AX,0            ;is offset zero?
      JNE   NEXT            ;if no, examine the next one
      CMP   BX,0            ;if yes, check for the seg value
      JNE   NEXT            ;if no, examine the next one
;displaying the interrupt number for the unused one
      .........
NEXT: LOOP  BACK            ;repeat until all entries are checked
```

Another approach is to code the program to run on a 386 or higher PC. The contents of the interrupt vector table for each interrupt entry are moved into EAX using 32-bit instructions and tested to see if they all zeros.

```
;assuming ES=0, SI=0 (pointers to interrupt vector table)
      .386                  ;386 instructions
BACK: MOV   EAX,ES:[SI]     ;get the seg:offset value
      ADD   SI,4            ;point to next entry
      COMP  EAX,0           ;is seg:offset zero?
      JNE   NEXT            ;if no, examine the next entry
;displaying the interrupt number for the unused one
      .........
NEXT: LOOP  BACK            ;repeat until all entries are checked
```

Activity 3

Compile and run the following C program which lists the unused interrupts of a PC. It shows how to access the interrupt vector table. Study the method and then proceed to Activity 4. Read Chapter 28 of the textbook about defining a far pointer and how to assign a fixed value to it.

```
/* This program lists unassigned interrupts by testing for zero entries of
the interrupt vector table */
        #include <stdio.h>
        #include <dos.h>
        main()
        {
        unsigned char far *vectable;          /* a FAR pointer */
        vectable=(unsigned int far*)=0;       /* make vector table address to 0 */
        unsigned int intnum;
        printf("The following interrupt numbers are unassigned\n");
        for(intnum=; intnum!=256; intnum++)
                {
                if(*vectable==0)
                printf("%d\t",intnum);
                ++vectable;
                }
        }
```

Activity 4

Study the C program in Activity 3, then write a C program to list the INT#, seg:offset of its handler, and its category as follows. The INT# and seg:offset address should be in hex format.

INT#	Handler Address Segment:Offset	Category

An interrupt can belong to one of the categories of (1) unused, (2) BIOS, or (3) DOS. The interrupt is assigned to BIOS if the segment address=F000H; otherwise, it belongs to DOS. In Assembly language it is easy to access the seg:offset address as was shown in Activity 2. However, in C language you need to use the C functions FP_SEG and FP_OFF. These two functions are designed specifically for 80x86-based PCs in order to read the segment and offset values. The following C code shows the way these two functions are used in the context of a C program for Activity 3.

```
segment = FP_SEG (*vectable);    /* get the seg value */
offset = FP_OFF (*vectable); /* get the offset value */
if(segment==0xF000)              /* is it BIOS ? */
        ...
```

When running the above program the following points should be noted:

1. If the entry for seg:offset value is 0000:00CF, there is no handler for this interrupt. CFH is the opcode for the IRET instruction. If this interrupt is activated, it simply reruns it without doing anything instead of hanging up the system.
2. If the entry for the segment portion of the seg:offset value is between C000H to EFFFFH, it must belong to an interrupt whose handler is provided by the ROM of the add-in card.
3. For interrupts 1DH, 1EH, 1FH, 41H, 43H, and 46H, the entry for seg:offset value is really a pointer value pointing to a table. For example, the entry for INT 41H is pointer pointing to the address of the hard disk parameter table.

See Question 4 in the worksheet.

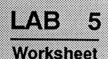

Name: _____
Date: _____
Class: _____

1. The hardware interrupts are referred to as _____ in MSD.

2. In Activity 1, compare the IRQ assignment as detected by MSD with Table 14-9 of the textbook. Are there any differences?

3. Using the MSD utility, make a table similar to Table 14-9 of the textbook for your PC.

4. In DEBUG, use the Dump command to dump the interrupt vector locations 0000:0000 - 0000:000F. Examine the seg:offset entry for INT 0 to INT 3. Do they match the result you get from running the program you created in Activity 4?

LAB 6

VIDEO PROGRAMMING

Objectives

» To code an Assembly language program to display letters in various colors

» To code a C program to display the video adapter type of the PC, video mode, and number of rows and columns per screen

References

Mazidi, Volumes I & II: Chapters 16 and 28

Materials

IBM PC (or compatible) PC with color monitor
MASM, TASM (or compatible) assembler
Turbo C compiler

Activity 1

Use INT 10H to code an Assembly language program to do the following in a PC with a VGA monitor:

1. clear the screen
2. set the cursor at row = 10, column = 6
3. display your name (first name followed by last name with a space in between), place of birth (town and state), giving each letter a different color. Use black as the color for the first letter, blue for the second one, green for the third one, and so on, all the way to high intensity white (see Table 16-3 in the textbook). If the number of letters is less than sixteen, fill in the rest with numbers 1, 2, 3, etc. until you use all 16 possible colors of Table 16-3.

Activity 2

After studying Chapter 28 of the textbook, use the "union REGS" method with INT 10H to write a C program to display the following information about the video monitor of your PC.

1. current video mode
2. active video page
3. number of columns in the screen
4. number of rows in the screen
5. type of video adapter

Notes: For parts 1, 2, and 3, you can use INT 10 function 0FH (see Appendix E.1 in the textbook). For part 4 you must use the BIOS data area location 0040:0084H (see Appendix H in the textbook). Also see Chapter 28 on how to access the BIOS data area using a C pointer. For part 5 you must use INT 10 option AH=1AH and AL=00. Before calling INT 10, set AX=1A00H. Not all PC's support this one (especially early PCs). Upon return, if AL=1AH it means that this function is supported by your BIOS. In that case, register BL has the adapter type as shown below:

```
BL=adapter type (only if AL=1AH)
00H = no display
01H = monochrome
02H = CGA
03H = reserved
04H = EGA (color)
05H = EGA (monochrome)
06H = PGA (Professional Graphics Controller)
07H = VGA (analog monochrome display)
08H = VGA (analog color display)
```

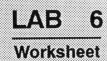
Use INT 10H for the following questions. Assemble and run the programs in DEBUG. For example, the code below shows how to get the current cursor position.

```
MOV AH,3
INT 10H
INT 3
```

1. Show the code to get the current video mode.

2. Show the code to set the video mode to mode 7.

3. Show the code to get the current cursor position.

4. Show the code to set the cursor to a new position.

5. Show the code to get the cursor's top and bottom scan lines.

LAB 7

EXPERIMENTING WITH KEYBOARD INPUT

Objectives

» To understand the role of options AH=0 and AH=1 of the INT 16H
  BIOS keyboard interrupt function call
» To write an Assembly language program to sound the speaker continu-
  ously while monitoring for a key press to stop the sound

References

Mazidi, Volumes I & II: Chapter 4

Materials

80x86 IBM (or compatible) computer
MASM (or compatible) assembler
Microsoft CodeView
Turbo C (or compatible C compiler)

Introduction

The major difference between the options AH=0 and AH=1 of INT
16H needs to be emphasized:

In the case of AH=1, BIOS monitors the keyboard to see if a key press
is available. If a key press is available, it sets ZF=0. If no key press is
available, it sets ZF=1 and returns without waiting for a key press. In other
words, AH=1 is used to monitor an event. This in is contrast to INT 21 option
AH=01 or AH=7 that wait for a key press and will not go on until a key is
pressed. When a key is pressed, ZF becomes 0, and we must use AH=0 of
INT 16H to get the pressed key.

Option AH=1 INT 16H is widely used where we need to monitor
some event while checking the keyboard to terminate the event.

Activity 1

Part (a)

Write a program in Assembly language to sound the bell (ASCII 07) continuously. The program should monitor the keyboard. When the Esc key is pressed, it should stop and go back to DOS.

Part (b)

Repeat Part (a), except check for letters "G" or "g" to stop sounding the bell.

Activity 2

The following are some equivalent functions for C and INT 16H on the 80x86 IBM PC and compatibles.

INT 16H	C functions
AH=0 or AH=10H	getche()
	getch()
AH=1 or AH=11H	kbhit()

Caution: In a given program, do not use INT 16H functions and C keyboard functions right after each other because it can cause problems for the keyboard buffer. It is recommended not even to use them in the same program.

Repeat Activity 1, Part (b) using C language.

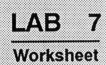
1. What are the contents of register AX upon return from INT 16H function call option AH=0?

2. What are the contents of register AX upon return from INT 16H function call option AH=1? What is the status of ZF and what does this indicate?

3. Which option of INT 16H is used to check for key press, AH=0 or AH=1?

4. Which option of INT 16H is used to get the character after a key press is detected, AH=0 or AH=1?

5. Getting the character always comes _____ (after, before) the key press detection.

LAB 8

EXPERIMENTING WITH HOT KEYS

Objectives

» To write an Assembly language program to play music upon pressing a hot key
» To write a C program to play music upon pressing a hot key

References

Mazidi, Volumes I & II: Chapters 10, 13, 18, 28

Materials

80x86 IBM (or compatible) computer
MASM (or compatible) assembler
Microsoft CodeView
Turbo C++

Activity 1

Write an Assembly language program to perform the following.
1. Prompt the user as follows, then get the user's response.

Enter ALT F5 for the music menu
Enter Q to go back to DOS

2. Upon selection of ALT F5 it should prompt the user as follows.

Enter Ctrl F1 to play the music
Enter Ctrl F3 to see the words to the song
Enter Ctrl F7 to go back to DOS

During the playing of the music the screen should state "Press Esc any time during the music to quit playing" and go back to DOS if Esc is pressed.

In the first two steps, if the user types any key other than what is indicated by the prompt the user should be reminded by stating a message such as "Try again" while displaying the prompts again.

For this activity you can modify Lab 4.

Activity 2

Complete Activity 1 using C language.

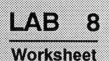

1. State the difference between the AH=0 option of BIOS INT 16H and the AH=01 (or AH=07) option of INT 21H.

2. State the difference between the AH=1 and AH=11H options of BIOS INT 16H.

3. State the difference between the AH=2 and AH=12H options of BIOS INT 16H. Which one did you use in the activities of this lab?

3. In order to perform full duplex, we must have how many data lines, and why?

LAB 9

LPT AND COM PORT INFORMATION

Objectives

» To examine the LPT and COM ports of a given PC

References

Mazidi, Volumes I & II: Chapters 17, 18, 28

Materials

80x86 IBM (or compatible) computer
MSD and DEBUG utilities
Turbo C Compiler

In this LAB we use MSD software to examine the LPT and COM ports for a given PC and then write a C program to verify the results of MSD.

Activity 1

Run the MSD (Microsoft Diagnostic) utility and examine the number of COM and LPT ports installed in your PC. Also find the base I/O port address assigned to each port.

Activity 2

Use DEBUG to dump memory locations 40:0 to 40:0F and verify the results in Activity 1.

Activity 3

Write a C program to display the base I/O address of the LPT and COM ports. If no port is installed it should state so:

COM 1: is installed with base I/O address of 3F8H
COM 2: not installed
and so on

1. From Activity 1, give the I/O base address of all the LPT and COM ports installed in your PC.

   I/O base address

   COM 1
   COM 2
   COM 3
   COM 4

   LPT 1
   ...

2. What baud rate are your COM 1 and COM 2 set to?

3. From Activity 2, give the I/O base address of all the LPT and COM ports installed in your PC.

   I/O base address

   COM 1
   COM 2
   COM 3
   COM 4

   LPT 1
   ...

4. Do the results of MSD and DEBUG in Activities 1 and 2 match? Compare the results with the data in Appendix G.

5. Does the result of Activity 3 match the result of Activities 1 and 2?

LAB 10

MOUSE PROGRAMMING

Objectives

»   To find all the information about the mouse in a given PC.

References

Mazidi, Volumes I & II: Chapter 5

Materials

80x86 IBM (or compatible) computer
MASM
Turbo C Compiler

Activity 1

Write a program to determine whether a mouse is installed in a given PC. If so it should display all the information pertaining to the mouse, such as the number of buttons, major and minor version of the mouse driver, mouse type (bus mouse, serial mouse), and the IRQ used by the mouse.

Activity 2

Write a program to divide the screen into four equal sections (boxes), each with a different background color. If the mouse is clicked with the left button in a given box, the name of the color of that section is displayed. Clicking the right button should exit the program.

1. Using MSD, verify the results of Activity 1.

2. Which option of INT 33H is used to get information about the mouse?

3. Which option of INT 33H is used to find the button press information?

LAB 11

EXPLORING FLOPPY/HARD DRIVES

Objectives

» To explore the cluster size of various disks
» To write a C program to calculate the total capacity of the disk and available free space

References

Mazidi, Volumes I & II: Chapter 19

Materials

80x86 IBM (or compatible) computer
MASM (or compatible) assembler
Microsoft CodeView

Activity 1

In the 80x86 IBM PC, the sector size is always 512 bytes but the size of the cluster varies among disks of various sizes. In some disks, the cluster is 2 sectors (1024 bytes) and in some it is 4 sectors. The cluster size is always a power of 2: 1, 2, 4, 8, and so on (see Table 1). The fact that a file of 1-byte size takes a minimum of 1 cluster must be emphasized. Look at the following.
1. Insert a diskette and use the DIR command to get its free space.
2. Use a word processor (or EDIT which comes with DOS) to type the following and save it. Then use the TYPE command to verify the contents of the file.
    ABCDEFG
3. Use the DIR command again to get the free space on the diskette.
4. Subtract the free space of steps 1 and 3. See Questions 1 and 2 on the worksheet.

Table 1: Sector and Cluster Size for Various Hard Disks

Drive Size (bytes)	Sector Size (Kbytes)	Cluster Size (bytes)	Sectors per Cluster
16M to 127 M	4	2048	2
128M to 255 M	8	4096	4
256M to 512M	16	8196	8

From "All about Hard-Disk Drives" by TJ Byers, *MicroComputer Journal,* September/October 1994.

The following exchange is from *PC Magazine*, September 13, 1994, page 361.

I recently upgraded my hard disk from a 250MB IDE drive to a 1.6GB SCSI-2 drive. When I transferred all my files to the new disk, I was shocked to see that the same files now occupied almost 300MB! Is this a function of how SCSI stores data on the disk? How can the same data take up more space on my new drive than on the originals?

Francisco Rosales
Chiapas, Mexico

PC MAGAZINE: This is a frequently asked question, since many users have been similarly surprised. The problem is that DOS stores files using clusters, and the larger the partition on your disk, the larger the cluster will be. Only one file can be stored in a cluster, so a file that contains only one character will still tie up an entire cluster.

Hard disk partitions up to 127MB use 2K clusters. In partitions between 128MB and 256MB, clusters are 4K in size. In partitions between 256MB to 512MB, clusters are 8K, and in partitions between 512MB to 1.024GB they are 16K. Between 1.024GB and 2.048GB, the clusters are a whopping 32K each. In every case, an "average" file will always waste half of a cluster, and this loss applies to the last cluster of every file throughout the partition.

Minimizing the total storage loss on the disk requires that you back up all the data, repartition the disk so that it has more partitions using smaller clusters, and then restore the files. Three partitions of about 500MB apiece (8K clusters) plus one smaller one may make most efficient use of the available space on your particular disk.

The moral of the story is that you should carefully plan how you will use your hard disk when upgrading your drive. It's far easier to partition the drive the way you want from the start than to repartition it that way later.

-- Alfred Poor

Activity 2

Repeat Activity 1 for a hard disk. See Question 3.

Activity 3

Although one can use BIOS INT 13H to access the disk (both floppy and hard disks) it is recommended not to do so since it can have disastrous consequences. For example, if you mistakenly write to a sector belonging to the FAT, you may lose portions of files. For this activity, we use INT 21H DOS function call AH=36H. It provides the disks parameters such as number of sectors per cluster, number of clusters, and so on. It is summarized below.

Call with: AH=36H
               DL= drive number (0=default, 1= A, 2= B, 3= C, etc.)

Upon return: AX=FFFFH if drive code is invalid
                otherwise:
                AX=number of sectors per cluster
                BX=number of available clusters
                CX=bytes per sector
                DX=total clusters per drive

For this activity, use the REGS union to write a C program to display the following for a given drive. See Questions 4 and 5.
1. The number of sectors per cluster
2. The cluster size in bytes and Kbytes
3. Total disk capacity in bytes, Kbytes, and Mbytes
4. Available disk space in bytes, Kbytes, and Mbytes

Notes from the Trenches

The following exchange is from *PC Magazine*, October 25, 1994, page 313.

PC MAGAZINE: In our issue of June 28, 1994, a reader asked about the discrepancy between the advertised capacity of his hard disks (245MB and 345MB) and their capacities as reported by his computer (only 234MB and 330MB), I explained that the problem was probably the difference between formatted and unformatted capacity.

Many sharp-eyed readers wrote in to offer a more precise explanation: Hard disk vendors use a different measuring stick than computers. Vendors treat *megabytes* as "millions of bytes," so a disk with 245,000,000 bytes of formatted capacity is called a "245MB" hard disk.

Computers don't count that way, however; binary beasts that they are, they count by powers of 2. A kilobyte is 1,024 bytes, not 1,000, and a megabyte is 1,024 kilobytes. So a megabyte to a computer is not 1,000,000 bytes, but rather 1,048,576, which is 2^{20}. Divide 245,000,000 bytes by 1,048,576 and you get 233.65--which, when rounded up, results in the 234MB capacity reported by the reader.

So the correct answer to the question is that both numbers are for the formatted capacities, and the 5 percent variation is simply caused by a difference in how *megabyte* is defined.

Jim Rogers of Vancouver, B.C., went one step further to show a third way that disk capacities are measured. He pointed out that the high-density 3.5-inch floppy disk is commonly referred to as a "1.44MB" disk. In fact, it contains 1,474,560 bytes (2 sides times 80 tracks times 18 sectors per track times 512 bytes per sector). Figuring 1,000,000 bytes per megabyte would give a capacity of 1.47MB, but 1,048,576 bytes per megabyte yields only 1.41MB. It turns out that for floppy disks, a megabyte is defined as 1,024,000 bytes, or 1,000K (where a kilobyte is defined as 1,024 bytes, or 2^{10}). Divide the 1,474,560 by 1,024,000 and you get the familiar 1.44MB result.

As Humpty Dumpty said in *Through the Looking Glass*, "When *I* use a word, it means just what I choose it to mean--neither more nor less." Obviously, the disk manufacturers have taken a page from Mr. Dumpty's book.

-- Alfred Poor

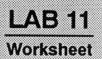
1. In Activity 1, what is the size of the file itself (the number of characters)?

2. In Activity 1, what is the cluster size in bytes? How many sectors are used for this cluster?

3. In Activity 2, what cluster size did you get?

4. Run the C program you created in Activity 3 to verify the cluster size for the floppy and hard disk used in Activities 1 and 2.

5. Run your C program from Activity 3 on various PCs with different hard disk sizes and make a table similar to Table 1.

LAB 12

FILE PROGRAMMING

Objectives

» To code an Assembly language program to create and write to a file
» To code an Assembly language program to read and display a file
» To write a C program to display file's date and time information

References

Mazidi, Volumes I & II: Chapters 19 and 28

Materials

80x86 IBM (or compatible) computer
MASM (or compatible) assembler
Microsoft CodeView

Activity 1

After studying Section 19.3 of the textbook, write an Assembly language program with following objectives. Then use the DOS TYPE command to dump the file onto the screen to verify its contents. In addition, copy it to the printer to get a hard copy.

1. Create a file on drive A.
2. Open it.
3. Get text from the keyboard using function INT 21H (AH=0AH).
4. Write it to the file.
5. Close the file.

Activity 2

Write an Assembly language program to read the file created in Activity 1 and dump it on the screen.

Activity 3

Using the REGS union, write a C program to get a file's date and time and display it on screen. In your program the user should be prompted for the file name. To get file's date and time, use DOS function call 21H option AH=57 hex. Notice that a file must be opened before we can get the file's date and time. See Appendix D for INT 21H DOS function calls.

1. When using a DOS function call in creating a file, how do we know if it was successful?

2. When is a file handle provided by the system?

3. When do we need to use a file handle?

4. What is ASCIIZ?

5. Why do we need to close a file after writing to it?

LAB 13

32-BIT PROGRAMMING

Objectives

» To examine the concept of scaled index addressing mode of the 386
» To examine the bit scan instruction of the 386 processor

References

Mazidi, Volumes I & II: Chapter 8, Section 21.1

Materials

386, 486, or Pentium IBM PC (or compatible) computer
MASM 5.1 or higher (or compatible) assembler
Microsoft CodeView or Borland Turbo Debugger

Activity 1

Using the 32-bit registers of a 386 core processor and scaled index addressing mode, write an Assembly language program to add a fixed value of 50000 to each element of following data array:

MYDATA_1   DD   10 DUP (2000000)

Use CodeView (or Turbo Debugger) to view and monitor the 32-bit registers' contents as you single-step through the program one instruction at a time. Remember that DEBUG will not let you view 32-bit registers of the 386. You can use your own array size and values.

Activity 2

Using the 32-bit registers of a 386 core processor and scaled index addressing mode, write an Assembly language program to find the total sum of all 10 elements of the following array of 32-bit data and save the result in memory.

MYDATA   DD   10 DUP (5000000)

Use CodeView (or Turbo Debugger) to view and monitor the 32-bit registers' contents as you single-step through the program one instruction at time. Use a calculator to verify the final sum provided by the CPU. Remember that DEBUG will not let you view 32-bit registers of the 386. You can use your own array size and values.

Activity 3

Write an Assembly language with following components.

1. Prompt the user for a single digit number between 0 and 9.
2. Get the number from the keyboard and verify that it is a digit between 1 and 9. If it is within the range, go to the next step; otherwise, display a message and ask the user to try again.
3. Using the bit-scan instruction, find the first "1" in the binary code of the number entered by the user. The first high should be found from both directions.
4. Display the location of the first 1 from each direction. The locations are numbered from D7 to D0.

For example, if the user types in 4, the first high is located at 2 when scanning from right to left since 4 is binary 00000100. However, if the user types in 9 we have 0 as the location for the first high when scanning from the right and 3 when scanning from the left since 9 is binary 00001001.

1. List the 386 registers that can be used in scaled index addressing mode.

2. The total sum of many doublewords requires data type _____ to be stored in memory.

3. For the 386 CPU, list the registers that can be used as pointers into the data segment.

4. Find the contents of ECX and EBX after execution of the following code. Verify it by Code View or Turbo Debugger.

```
MOV   EAX,200000
BSF   ECX,EAX
BSR   EBX,EAX
```

LAB 14

MEM COMMAND AND DOS MEMORY MANAGEMENT

Objectives

» To understand and experiment with DOS's MEM command
» To experiment with memory management capabilities of DOS

References

Mazidi, Volumes I & II: Chapter 25

Materials

DOS 6.x
386, 486, or Pentium IBM PC (or compatible) computer
DOS EDIT (or any ASCII editor)

The MEM command in DOS 6.x

The MEM command is a very useful DOS command that allows us to examine the memory allocation of the PC. It has many options that are selected with switches. Next we describe various aspects of the MEM command.

In DOS 6.x, the MEM command with no additional parameters (switches) displays the total, used, and free memory for all types of PC memory such as conventional, upper, extended, and expanded. It also shows a category of memory called Adapter RAM/ROM, which is the amount of RAM beyond 640K that can be converted to upper memory with the use of the DOS=UMB statement in CONFIG.SYS. For example, if a PC with 386 and higher processors has 1M of RAM, 640KB is used for conventional memory and the rest (384K =1024 - 640) is categorized as Adapter RAM/ROM since it can be mapped into the address space A0000 - FFFFF which is normally used for Video RAM and adapter ROM. If in the CONFIG.SYS we have DOS=UMB, the Adapter RAM/ROM is reduced and added to the upper memory category, thereby allowing it to be used as upper memory block.

Note: Depending on the contents of the CONFIG.SYS file, the 384K bytes of RAM (1024 - 640 = 384) can be listed under the extended memory category instead of Adapter RAM/ROM.

Below are listed some of the most useful parameters of the MEM command.

MEM /C

While MEM with no parameter displays the amount of memory used in conventional memory, it does not indicate how it is used. We can use MEM /C (or MEM /CLASSIFY) to find out who uses the memory and how much.

Note: Since MEM /C displays more than one full screen, you will need to use MEM /C /P to pause in between each screen.

MEM /D

To see the exact memory location where each TSR and device driver is loaded, we use MEM /D (or MEM /DEBUG). This provides much more detailed information about memory usage than MEM /C.

MEM /F

To see which areas of conventional and upper memories are free, we can use MEM /F (or MEM /FREE). This option is helpful in finding the largest free upper memory block so that we can move a given device driver or TSR into it.

MEM /M

MEM /M (or MEM /MODULE) allows one to examine how a given module uses memory. This can be used to find out exact memory regions used by various programs, TSRs, and device drivers. To do that, the name of the module must be typed after the MEM /M command. For example, if DOSKEY is loaded we can examine its memory usage with MEM/M as follows. The second command would show how MSDOS uses memory.
MEM /M DOSKEY
MEM/M MSDOS

MEM command for DOS 5.0.

In DOS 5 the MEM command displays memory usage in a different format. For example, for MEM with no parameters, it displays the total conventional memory, bytes available to DOS, and the largest executable program size (all in bytes). In addition it also provides the total contiguous extended memory (memory beyond 1M) and the amount available (both in bytes). DOS 5 supports MEM /C. It also has MEM /P (or MEM/PRO-GRAM), which displays information concerning the system's memory and programs that use it plus the physical address of where they are located. DOS 5.0 does not support MEM /D and MEM /F commands. Notice that MEM /P in DOS 5 and DOS 6.x have different meanings.

Activity 1

In the DOS directory, type the following commands. Including PRN will send the result to the printer to get a hard copy of the output. Use a red pen or highlighter to mark and separate the results of each command on the printer output. Then analyze each output for your PC. See questions 1 and 6 on the worksheet.

```
C:\DOS>MEM  PRN
C:\DOS>MEM /C  PRN
C:\DOS>MEM /D  PRN
C:\DOS>MEM /F  PRN
C:\DOS>MEM /M MSDOS  PRN
C:\DOS>MEM /M DOSKEY  PRN
```

Activity 2

It is always necessary to keep a BOOT DISKETTE. This way, if your hard drive fails to boot you can start up your computer from a diskette drive. This is especially true when you are experimenting with the memory manager capabilities of MS DOS as we plan to do in this activity. The following are the steps to make the boot diskette.

1. Saving the CONFIG.SYS file
a. Format a new disk as a bootable disk as follows (in the DOS directory).
   C:\DOS>FORMAT A:/S
b. Copy your CONFIG.SYS as follows (in the main directory).
   C:\COPY CONFIG.SYS A:
c. You also need to copy any driver files your CONFIG.SYS uses. For example, if it uses ANSI.SYS you need to copy it as follows.
   COPY C:\ANSI.SYS A:

2. Saving the AUTOEXEC.BAT file
a. Copy your AUTOEXEC.BAT as follows (in the main directory).
   C:\COPY AUTOEXEC.BAT A:
b. You also need to copy any driver files your AUTOEXEC.BAT files uses.
   Note: It is also recommended to make a hard copy of the CONFIG.SYS and AUTOEXEC.BAT files and save it somewhere safe. This is done as follows.
   C:\DOS>TYPE CONFIG.SYS PRN
   C:\DOS>TYPE AUTOEXEC.BAT PRN

3. Saving the CMOS RAM. Make a copy of your CMOS RAM and save it. Major commercial utilities such as Norton Utilities and PC Tools let you do that. There are also some shareware software packages from electronic bulletin board services (BBS) which let you copy the CMOS RAM and save it. Whether or not you have access to any of this software, you should always make a hard copy of your system CMOS RAM and save it somewhere safe. To get a hard copy of your CMOS RAM contents do the following.
a. Turn your printer on.
b. Turn your PC on and during the power-on-self test (POST), hold down the DEL key. The screen indicates to enter a key (normally F1) to access your CMOS setup.
c. Print each screen of the CMOS setup (using Shift+PrintScreen).

Now that you have an emergency diskette, go to your CONFIG.SYS file and put the REM statements in front of the commands which are related to DOS memory management in order to block them and then reboot your system. In each case, use the MEM command or MEM /C/P to see the effects of various DOS memory management commands such as HIMEM.SYS, HIGH, and so on. Make a hard copy of each case and analyze and discuss the effects on the size of conventional memory, upper memory, and extended memory as you place REM or as you remove the REM (do not forget to reboot the system each time). Also discuss the effect on the largest executable program size and largest free upper memory block.

1. In Activity 1, how many K bytes is your installed conventional memory?

2. In Activity 1, how many K bytes is your memory beyond 1M address space?

3. In Activity 1, how many K bytes is the total memory for the UPPER and RE-SERVED categories?

4. In Activity 1, are any of your devices or TSRs loaded into upper memory? If yes, which ones? If no, explain why in terms of your CONFIG.SYS data.

5. In some PCs the number of bytes available to DOS is 1 or 2 Kbytes less than the size of installed conventional memory. Why? *Hint:* See the discussion of extended BIOS data area in Section 14.2 of the textbook. In Activity 1 compare the size of total conventional memory and the number of bytes available to DOS. Are they equal on your PC?

6. Why is the largest executable program size smaller than the number of available bytes to DOS?

LAB 15

WRITING AND TESTING TSR PROGRAMS

Objectives

» To examine conventional memory usage by TSR programs
» To write and test a TSR program using BIOS INT 17H

References

Mazidi, Volumes I & II: Chapters 24, 25 and Section 18.2

Materials

80x86 IBM (or compatible) computer
MASM (or compatible) assembler
Microsoft CodeView

Activity 1

This activity shows the memory usage by a TSR. Perform the following. Then answer questions 1 and 2 on the worksheet.

1. Get a hard copy of the MEM output using the DOS command as follows.
   C:\DOS>MEM /C PRN
2. Assemble and link TSR Program 24-1 of the textbook.
3. Using the EXE2BIN utility, convert it to a COM file and call it MYTSR.COM. Save it on a floppy disk.
4. Run MYTSR.COM. This will load the TSR.
5. Test MYTSR to see if it is working by pressing ALT and F10 at the same time. You should hear a beep.
6. In the DOS directory, repeat step 1. You should see MYTSR.COM as one of the programs using a portion of the 640K conventional memory. Examine MYTSR's conventional memory usage by comparing the output from this step and step 1.
7. To unload MYTSR from conventional memory, reboot the system. There are programs available through many electronic bulletin board services (BBS) which allow you to unload a TSR without rebooting the system.

Activity 2

Write and test a TSR program to print your last name every time you press SHIFT F7. Use BIOS INT 17H to print characters. Also in your TSR, send 0CH as last character so you can see if your TSR is working properly (0CH is ASCII for Form Feed to eject a page from the printer). For a discussion of INT 17H, see Chapter 18 of the textbook. Verify the loading of your TSR as shown in Activity 1.

Note: Testing a TSR requires repeated trying and rebooting the PC.

Name: _____

Date: _____

Class: _____

1. In Activity 1 step 1, what is the size of the largest executable program?

2. From step 6 in Activity 1, how many bytes of conventional memory are used by MYTSR?

3. In Activity 2, how many bytes of conventional memory are used by the TSR?

4. In Activity 2, what would happen if no form feed were sent as the last character?

5. Why do we avoid using DOS function calls in a TSR?

SECTION 4

Hardware Interfacing Via Expansion Slots

This section explores the interfacing of the PC to devices such as LCDs, stepper motors, keypads, DAC and ADC devices, sensors, and more. The labs in this section require the use of the PC Bus Extender card and PC Interface Trainer which are available from the source indicated below. In the absence of the PC Bus Extender card, you must wire-wrap them as shown in Appendices C and D of this lab manual.

Electronix Express (A Division of R.S.R. Electronics, Inc.)
365 Blair Road, Avenel, New Jersey 07001, USA
Phone: 1-800-972-2225   (In NJ 1-732-381-8020)
Fax: 1-732-381-1572   http://www.elexp.com

LAB 1

EXAMINING AND TESTING THE PC INTERFACE TRAINER

Objectives

» To examine and test the I/O ports of the PC interface Trainer

References

Mazidi, Volumes I & II: Chapter 12

Materials

PC Interface Trainer
Logic probe
DIP switches with pull-up resistors

As discussed in Section 12.7 of the textbook, the PC Interface Trainer is an easy-to-use and affordable module designed to be used for I/O interfacing in the x86 PC. To use this Trainer, you will also need the PC Bus Extender discussed in Section 12.7 of the textbook. To get the Trainer and Bus Extender you have two options:

1. Wire wrap them. Details of this process are given in Appendices B - D of this lab manual. Notice that if you decided to wire wrap the Bus Extender, you will need an 8-bit PC/XT Interface card with nothing on it. This card alone can cost you $30 to $40. See Appendix C.

2. To purchase them (assembled and parts kit) from the supplier listed on the first page of this section.

We will use this Trainer in all the labs in this section. The PC Trainer used is also referred to as the PC Interface Module since it is detachable via the cable.

Activity 1

Test the operation of the 8255 as follows.

1. Connect each bit of port A (PA0 - PA7) of the 8255 to an LED of your digital trainer, or you can use a logic probe.
2. Go to DEBUG (in the DOS directory) and program the 8255's control register to configure all three ports A, B, and C as output ports (control word 80H). Test the operation of the 8255 with the following code in DEBUG.

```
-A 100
        MOV   AL,80H        ;all ports as output (do not type H in DEBUG)
        MOV   DX,303H       ;control reg address
        OUT   DX,AL         ;send control word
        MOV   DX,300H       ;port A address
        MOV   AL,99H        ;send 10011001 data to LEDs
        OUT   DX,AL         ;send it to port A
        INT   3            ;stop
```

3. Run the above program by typing the go command as shown below. After execution, 10011001 should appear on the LEDs of the digital trainer.

```
-g=100 {return}
```

4. To detect if any bit is stuck to 0 or 1, replace the line "MOV AL,99" with "MOV AL,55" and run it again. Repeat the process to output AAH.

Repeat all the above steps separately for Ports B and C to ensure that all the ports work for output.

Activity 2

Test the 8255 for input port operation. This is done as follows.

1. Connect each bit of Port A (PA0 - PA7) of the 8255 to buffered DIP switches of your digital trainer.
2. Go to DEBUG (in the DOS directory) and program the 8255's control register to configure all three ports of A, B, and C as input ports (control word 9BH). Test the operation of the 8255 input ports with the following code in DEBUG.

```
-A 100

        MOV   AL,9BH        ;all ports as input (do not type H in DEBUG)
        MOV   DX,303H       ;control reg address
        OUT   DX,AL         ;send control word
        MOV   DX,300H       ;port A address
        IN    AL,DX         ;get the status of DIP switches
        INT   3            ;stop
```

3. Run the above program by typing the go command as shown below. After execution, examine the AL register on the PC screen to see if the bits on the DIP switch were input into AL.

   -g=100 {return}

4. To see if any of input lines are stuck to 0 or 1, set the DIP switches to 55H (01010101 binary) and run the DEBUG program again. Examine the AL register on the PC screen. Repeat the process by setting the DIP switches to AAH (10101010 binary).

   Repeat all the above steps separately for Ports B and C to ensure that all the ports work for inputting data.

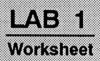
Examine the PC Interface Trainer and answer the following questions

1. Give the Port addresses used for the 8255.

2. Give the port address used for

   PA =

   PB =

   PC =

   CNTR Reg =

3. Indicate which header (H1, H2, ...) brings the bus signals from the PC to the Trainer.

4. What is the Role of H2 in the PC Interface Trainer?

5. List all the headers that are dual row.

6. In dual-row headers _____(odd, even) pins are ground signals.

7. Which header allows you to take the ports A, B, C of the 8255 off the board with a 50-pin cable?

LAB 2

SIMPLE I/O WITH THE 8255 OF THE PC INTERFACE TRAINER

Objectives

» To perform I/O operations with ports A and B of the 8255 of the PC Interface Trainer

References

Mazidi, Volumes I & II: Chapter 12
Appendices B and D of this lab manual

Materials

PC Interface Trainer
Logic Probe
DIP Switches with pull-up resistors.

In this lab we need to set up Port A of the 8255 as input and Port B as output. In this manner, we bring in 8-bit binary data and send it out to 8 LED indicators. To do that, connect all bits of port A to DIP switches of your digital trainer and all bits of port B to LEDs of the digital trainer (If you do not have access to a digital trainer, use DIP switches with 10K pull-up resistor and for LEDs use a logic probe). Run a simple test in DEBUG to make sure the set-up works (See Lab 1 of this section or Section 12.7 in the text book). This set-up will be used in the next three activities.

Activity 1

Part (a)

After you set up Ports A and B as discussed above, write a program to get the data from Port A and send it to Port B. In other words, 8 LEDs always reflect the status of the 8 DIP SW. Change the status of SW and run the program each time. Repeat this several times.

Part (b)

After making sure that Part A works, modify your program so that the LEDs reflect the status of SW continuously. When any key (or ESC key) is pressed, it should exit and go back to DOS. Notice that you need to scan the keyboard (using INT 16H, AH=01) before the status of SW is scanned.

Activity 2

Using the setup in Activity 1, write a program (in C or Assembly) to do the following.

1. Prompt the user for a character.
2. Get the character using INT 21H with function AH=01.
3. Check the character. If it is a letter (A - Z, a - z) or a number (0 - 9), it is sent to the output port to be displayed on the LEDs; otherwise, the user is asked to try again. The LEDs should have all zeros when no character is sent to the output port.

Reminder: Characters displayed on LEDs are in ASCII (e.g., character 'A' is shown as 41H or 010000001 binary).

Activity 3

Write a program that continuously gets a binary number from Port A, converts it to ASCII, and displays the result on the PC screen. Pressing ESC should exit the program. Change the binary numbers on the switches and observe the result on screen. Verify the result with a calculator.

Note: This is 8-bit binary (hex) to decimal (and then to ASCII) conversion. ASCII data is displayed on the screen.

Notes from the Trenches

In some PCs, the use of two "OUT" or two "IN" instructions right after each other will not work, as in the following code.

```
MOV   AL,55H
OUT   PORT_A,AL
OUT   PORT_B,AL
```

In the absence of some other instructions in between subsequent OUT or IN instructions, you need to put several "NOP" instructions (or a time delay, as preferred by Intel) in between them. For example the above program could be coded as follows.

```
MOV   AL,55H
OUT   PORT_A,AL
NOP
NOP
NOP
NOP
OUT   PORT_B,AL
```

The same is true for an "OUT" instruction followed by "IN" or an "IN" instruction followed by an "OUT".

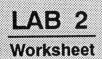
Name: _____
Date: _____
Class: _____

1. For Activity 1, which I/O mode of 8255 did you use?

2. In Activity 2, if the user types in 'S', give the status of the LEDs.

3. In Activity 2, if the user types in '9', give the status of the LEDs.

4. For Activity 3, if the switches have 11111101 (LSB on the right), what do we see on the PC screen?

5. For Activity 3, to see 198 displayed on the screen, what should be the status of the switches?

6. Assume that register AL= F9 hex. Write a program to convert this hex (binary) data to ASCII. Do not use any type of jump instruction. At the end of the program, registers DL, DH, CL each should have one of the ASCII digits.

LAB 3

INTERFACING A PRINTER TO A PC

Objectives

» To interface a Centronics printer to the 8255

» To gain an understanding of printer handshaking signals

References

Mazidi, Volumes I & II: Chapters 12 and 18

Materials

80x86 IBM (or compatible) computer
MASM (or compatible) assembler
MicroSoft CodeView
PC Interface Trainer
Centronics printer cable

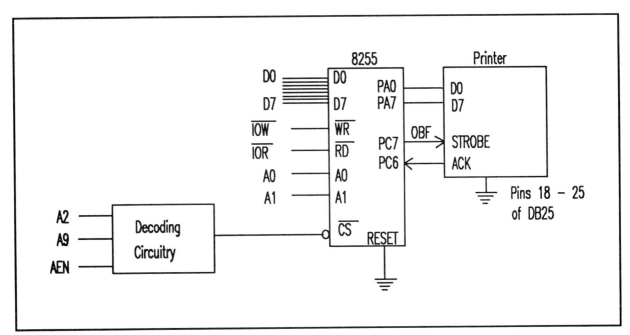

Figure 1. 8255 Connection to Printer

Activity 1

Examine the Centronics cable pins listed in Chapter 18 of the textbook. Connect the data bus and handshaking signals of the printer to your 8255-to-PC interface as shown in Figure 1. The signal connection is according to Example 12-9 of the textbook. Assemble and run the program shown in that example and verify its operation.

Activity 2

Make the following modifications in the program of Activity 1.

In the data segment, put your name (last name, first name), date, time, course number and section, and the objective of the lab. Each of the above must be printed on a separate line with an extra line in between. Also put 0CH (ASCII code for form feed) right before the '$' delimiter to eject the printed page.

Note: In this lab, make sure that the GND signals of the parallel port (pins 18 - 25 on DB-25) are also connected to a ground point on your 8255-to-PC interface module.

1. Indicate the direction of the following signals from the printer's point of view.

Signal	Direction
D0-D7	
STROBE	
ACK	

2. Indicate the direction of the following signals from the 8255's point of view.

Signal	Direction
PA0-PA7	
OBFa	
ACKa	

3. For each of the following signals, indicate its normal status (high or low) and also indicate the activation level for each (active low or active high).

Printer's Handshake Signal	Normal State	Active State
Strobe		
ACK		

8255 Handshake Signals	Normal State	Active State
OBF		
ACK		

LAB 4

INTERFACING AN LCD TO THE PC

Objectives

» To understand the operation modes of an LCD (liquid crystal display)
» To interface and program an LCD to an 8255

References

Mazidi, Volumes I & II: Chapter 12
Dot Matrix LCD Module: Character-type DMC Series User's Manual by Optrex Corp.

Materials

80x86 IBM (or compatible) computer
MASM (or compatible) assembler
MicroSoft CodeView
PC Interface Trainer
20x2 LCD DMC20261 from Optrex DMC series or a compatible one

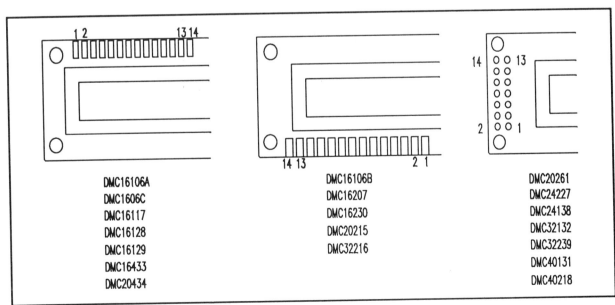

DMC16106A	DMC16106B	DMC20261
DMC1606C	DMC16207	DMC24227
DMC16117	DMC16230	DMC24138
DMC16128	DMC20215	DMC32132
DMC16129	DMC32216	DMC32239
DMC16433		DMC40131
DMC20434		DMC40218

Figure 1: Pin Positions for Various LCDs from Optrex Corporation

Activity 1

Connect the LCD to your PC Interface Trainer as shown in Figure 2. Then write and run a program to display your name on line 1 of the LCD (first name followed by last name with a space in between).

Note: If you are not monitoring the busy flag of the LCD, you will need a 20 ms delay in your program. The delay must be independent of the microprocessor speed. For x86 PCs with 286 and higher CPUs, use the following delay routine.

```
;(CX) = COUNT OF 15.085 MICROSECOND
           WAITF      PROC NEAR
                      PUSH AX
           WAITF1:

                      IN    AL,61H
                      AND   AL,10H      ;CHECK PB4
                      CMP   AL,AH       ;DID IT JUST CHANGE
                      JE    WAITF1      ;WAIT FOR CHANGE
                      MOV   AH,AL       ;SAVE THE NEW PB4 STATUS
                      LOOP  WAITF1      ;CONTINUE UNTIL CX BECOME
                      POP   AX
                      RET
           WAITF      ENDP
```

Activity 2

Repeat Activity 1 while also putting the year you graduated from high school on the second line. When you run your program the LCD should show (for example):

Alex Young
Graduated in 1978

Activity 3

Write an LCD program to display your last name on the first line and the current year on the second line. Both should be in the middle of the line.

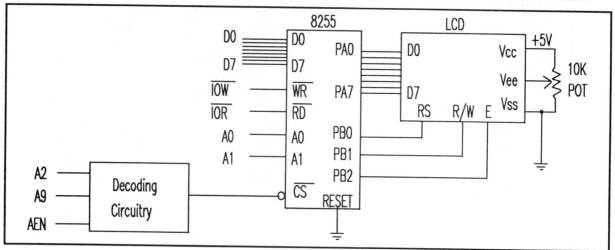

Figure 2: LCD Connection to the 8255

1. How does the LCD distinguish data from instruction codes when receiving information at its data pin?

2. To send the instruction code 01 for the clear display, we must make RS = ___.

3. To send letter 'A' to be displayed on the LCD, we must make RS = ____.

4. What is the purpose of the E line? Is it an input or an output as far as the LCD is concerned?

5. When is the information (code or data) on the LCD pin latched into the LCD?

6. What is the purpose of the BF (busy flag)? Describe how is it accessed.

LAB 5

INTERFACING A STEPPER MOTOR TO A PC

Objectives

» To interface a stepper motor to the 8255
» To write a program in which the angle and direction of stepper motor rotation is controlled from the PC keyboard

References

Mazidi, Volumes I & II: Chapter 12
Design Engineer's Guide to DC Stepping Motor, Superior Electric Company
The Handbook of Personal Computer Instrumentation, 5th ed., Burr-Brown, 1990

Materials

80x86 IBM (or compatible) computer
MASM (or compatible) assembler
PC Interface Trainer
Stepper motor
ULN2003 driver chip

Activity 1

Using your 8255-to-PC Interface Trainer, build the circuit shown in Figure 2 on the breadboard. The following steps show the 8255 connection to the stepper motor and its programming.

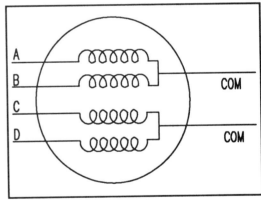

Figure 1. Stator Windings

Table 1: Normal 4-Step Sequence

Clockwise	Step #	Winding A	Winding B	Winding C	Winding D	Counter Clockwise
↓	1	1	0	0	1	↑
	2	1	1	0	0	
	3	0	1	1	0	
	4	0	0	1	1	

Note: 0=off and 1=on

1. Use an ohmmeter to measure the resistance of leads. This should identify the COM leads and A through D winding leads.
2. The common wire(s) are connected to the positive side of the motor's power supply. In many motors, +5V is sufficient.
3. The four leads of the stator winding are controlled by four bits of the 8255 port (in this case PA0 - PA3). However, since the 8255 lacks sufficient current to drive the stepper motor windings, we must use a driver such as the ULN2003 to energize the stator. Instead of the ULN2003, we could have used transistors as drivers, as shown in Figure 3. However, notice that if transistors are used as drivers, we must also use diodes to take care of inductive current generated when the coil is turned off. One reason that the ULN2003 is preferable to the use of transistors as drivers is the fact that the ULN2003 has an internal diode to take care of back EMF.
4. In DEBUG, write and run the following program. In this program, the stepper motor keeps rotating until you press any key on the IBM PC keyboard to stop it. *Reminder*: DEBUG assumes that all numbers are in hex. Do not type the comments; they are given for sake of clarification.

```
-A 100
        MOV  AL,80       ;control word for all 8255 ports as out
        MOV  DX,303      ;control reg address
        OUT  DX,AL       ;to control reg
        MOV  BL,66
        MOV  AH,01       ;check the key press
        INT  16          ;using INT 16
        JNZ  011D        ;go to stop if any key pressed
        MOV  AL,BL       ;otherwise send pulse to stepper motor
        MOV  DX,300      ;port A address
        OUT  DX,AL
        MOV  CX,0FFFF    ;(change this value to see rotation speed)
        LOOP 0117        ;wait for delay
        ROR  BL,1        ;rotate for next step
        JMP  108         ;and continue until a key is pressed
        INT 3            ;stop
```

5. Run the above program with g=100 and see the motor keep rotating on and on. By pressing any key, it will stop.
6. Change the count value of CX for various delays to see the speed of rotation.

Table 2: Stepper Motor Step Angles

Step Angle (Degree)	Steps Per Revolution
0.72	500
1.8	200
2	180
2.5	144
5	72
7.5	48
15	24

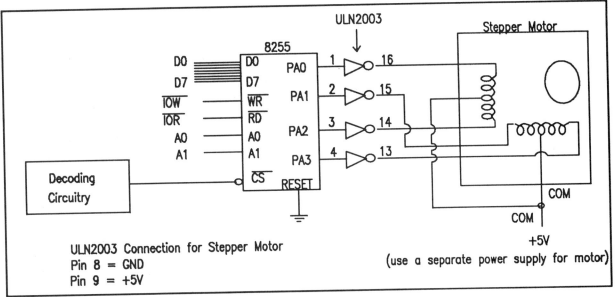

Figure 2. 8255 Connection to Stepper Motor (using ULN2003)

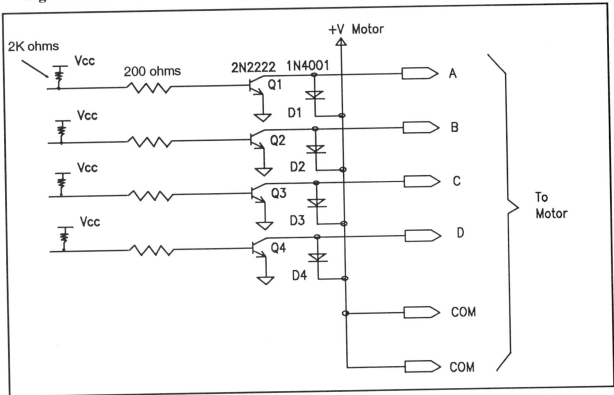

Figure 3. Using Transistors for Stepper Motor Driver (Modified from Burr-Brown)

Notes from the Trenches

The ULN 2003 IC is an excellent driver for stepper motors. In the absence of that you can use NPN transistors (see Figure 3). Remember that the ULN 2003 can provide a maximum of 600 mA. If you need current higher than that, use a TIP 110 (or TIP 120) transistor. The base resistors normally have a 10-to-1 ratio. The exact value can vary from motor to motor, depending on the current needed for a given motor.

Activity 2

After making sure that your stepper motor connection in Activity 1 works, write and run a program with following components.

1. Prompt the user for the direction of motor rotation, where R=clockwise and L=counter clockwise (or 1=CW and 2=CCW, or any other format).
2. After user has made a choice, the motor will rotate accordingly.
3. The motor should stop when the user presses Esc on the IBM PC keyboard.
4. In your program, make sure to use a fixed time delay as discussed in Chapter 13 of the textbook. The time delay in between each motor sequence must not depend on the speed of the CPU. The loop time delay used in Activity 1 is CPU dependent. Also, make sure that your time delay is long enough.

Activity 3

After making sure that your hardware setup in Activity 1 works, write and run a program with the following components. The user is prompted for:

1. The direction of motor rotation (clockwise or counter clockwise).
2. The angle of rotation (45, 90, 180, 270, 360, or whatever degree you want).

Activity 4

After making sure your setup in Activity 1 works, write and run a program with the following components. The user is prompted for:

1. The number of steps for the stepper motor to rotate.
2. The direction of motor rotation.

1. What is a step angle? Define steps per revolution.

2. If a given stepper motor has a step angle of 5 degrees, find the number of steps per revolution.

3. In Activity 1, why are we using 66H (01100110 binary) instead of 6 (0110 binary)?

4. Using the ROL instruction, show all the four-step sequences if the initial step is 1001 (binary).

5. Give the number of times the four-step sequence in Table 1 must be applied to a stepper motor to make a 100-degree move if the motor has a 5-degree step angle. Also fill in the characteristics for your motor below.

   Step angle   _____
   Steps per revolution   _____
   Number of rotor teeth   _____
   Degree of movement per 4-step sequence   _____

LAB 6

INTERFACING A 5X4 KEYBOARD TO A PC

Objective

» To interface a 5x4 keyboard (keypad) to a PC

Reference

Mazidi, Volumes I & II: Chapter 18

Materials

80x86 IBM (or compatible) computer
MASM (or compatible) assembler
MicroSoft CodeView
PC Interface Trainer
5x4 keyboard (part 86JB2-203 from Grayhill) or any N x M matrix keyboard
9 of 6.8K ohms resistor

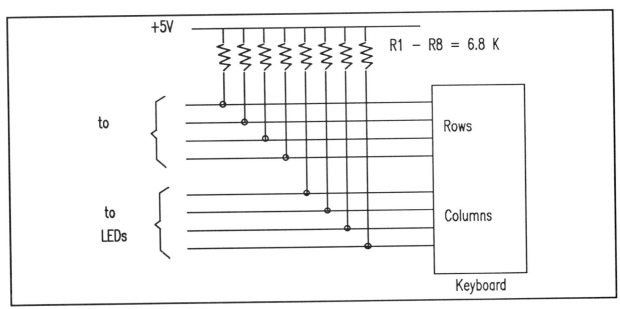

Figure 1. Testing Keyboard Matrix

In many small projects the use of a keyboard as an input device is unavoidable. In this lab we discuss the 5x4 matrix keyboard and then show how to interface it with a microprocessor. Although the keyboard that we use from Grayhill (5x4 with part number 86JB2-203) is a pricey one, you can replace it with any other matrix keyboard.

Activity 1

The first step is to make a truth table for the keyboard. The truth table shows the row and column contacts by which a key is produced. Connect the ohmmeter leads, one to a row and one to a column terminal (lead) of the keyboard, and press the keys one at a time until you measure zero ohms. Repeat the process until the all the keys are mapped.

Activity 2

To verify the truth table generated in Activity 1, follow these steps.

1. Connect the rows of the keyboard to switches of your digital trainer as shown in Figure 1.
2. Connect the columns of the keyboard to the LEDs of your digital trainer as shown in Figure 1. Notice that after this step all the LEDs are "on" if they are connected properly.
3. Ground the first row while all other rows are high. Then press each key in that row and examine the status of the LEDs. Only one of them must be "off" as you press a given key in that row. This should match your truth table of Activity 1.
4. Ground the second row while all other rows are high. Then press each key in that row and examine the status of the LEDs.
5. Repeat step 4 for all rows to verify the complete mapping of your keyboard's truth table.

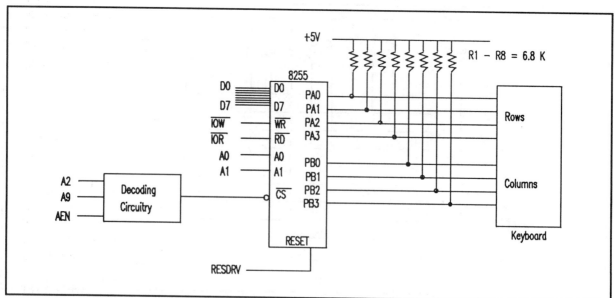

Figure 2. 8255 Connection to Keyboard

Activity 3

Go through the following steps to connect and test your keyboard interfacing to the PC via the 8255 of the PC Interface trainer.

1. Connect your keyboard to Ports A and B as shown in Figure 2. Port A is for the rows and Port B is for the columns.
2. In DEBUG, code the following program. Remember that DEBUG assumes that all numbers are in hex.

```
-A 100
        MOV   AL,82H      ;PA=OUT,PB=IN
        MOV   DX,303H     ;control reg port address
        OUT   DX,AL
A1:     MOV   AH,01       ;check the PC keyboard to stop this program
        INT   16H         ;using INT 16H BIOS
        JNZ   A3          ;get out if any key pressed from PC keyboard
        MOV   DX,300H     ;port A address
        MOV   AL,0        ;ground all rows
        OUT   DX,AL       ;send it to port A
        MOV   CX,200H     ;wait for a delay
A2:     LOOP  A2
        MOV   DX,301H     ;port B address to read the columns
        IN    AL,DX       ;get the character from our own keyboard
        AND   AL,0FH      ;mask the upper four bits
        MOV   AH,02       ;AH=2 of INT 21H to display char on PC monitor
        MOV   DL,AL
        INT   21H         ;display char from our keyboard on PC monitor
        JMP   A1          ;keep repeating until PC keyboard is pressed
A3:     INT   3           ;to get out of this loop
```

Now run the program by making G=100. If your hardware is connected properly, as you press any of the keys in a given column it will display its equivalent ASCII character on the PC monitor. Since your keyboard is 4x4, it should display the ASCII characters associated with 07, 0B, 0D, and 0E (all hex). If no key is pressed, the character for 0FH is displayed continuously. For the characters, see Appendix F of the textbook. Notice that for characters such as BELL (07), the PC will beep.

Activity 4

After making sure that your hardware setup in Activity 3 works properly, write and run a program that scans your keyboard and displays on the PC monitor any character pressed by the user. Your program must display keys 0 - 9 as numbers 0 - 9 on the PC screen while keys 10, 11, 12, 13, 14, 15 (or you might say 0A - 0F in hex) are displayed as letters A, B, C, D, E, and F, respectively. The choice of characters associated with key numbers beyond 0F is up to you but do not make Esc one of them since pressing the Esc key on the PC keyboard should get your program out of the loop and back to DOS.

You can modify and incorporate Program 18-1 of the textbook for your program. You can also use the 80x86's XLAT instruction for the table conversion.

1. What is the purpose of generating the truth table for a given keyboard?

2. What is the purpose of grounding each row in keyboard interfacing?

3. In Figure 1, what is the status of the LEDs if no key is pressed?

4. In Figure 1, what is the status of the LEDs if all the rows are grounded?

5. True or false. In an N x M matrix keyboard such as shown in Figure 1, we cannot press two keys at the same time.

6. In your program in Activity 4, how is the key press detected?

7. In your program in Activity 4, how is a key press identified?

LAB 7

DIGITAL-TO-ANALOG CONVERTER (DAC) INTERFACING TO A PC VIA EXPANSION SLOT

Objectives

» To interface a DAC to the 8255
» To generate a sine wave on the scope using the DAC

References

Mazidi, Volumes I & II: Chapter 12
Linear and Interface Integrated Circuits, Motorola Corp.
Data Acquisition Linear Devices Databook, National Semiconductor Corp.

Materials

80x86 IBM (or compatible) computer
MASM (or compatible) assembler
PC Interface Trainer
MC1408 DAC (or its equivalent DAC0808)
100 pf capacitor
Resistors: 2 of 1K, 2 of 1.5K
5 K potentiometer

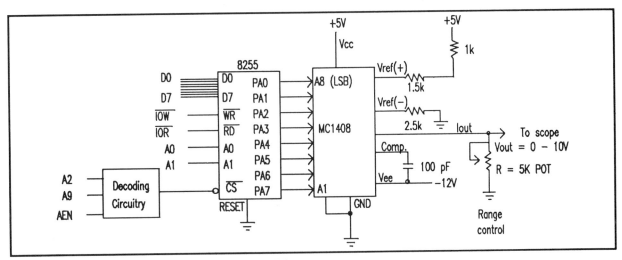

Figure 1. DAC Connection to the 8255

Activity 1

In order to generate a stair-step ramp, set up the circuit in Figure 1 and connect the output to an oscilloscope. Then assemble and run the following program.

```
            MOV   AL,80H      ;all ports as output
            MOV   DX,303      ;control reg address (change it for your design)
            OUT   DX,AL
A1:         MOV   AH,01       ;check for key press
            INT   16H         ;using PC BIOS INT 16
            JNZ   STOP        ;stop if any key is pressed
            SUB   AL,AL       ;other wise generate a stair-step ramp
            MOV   DX,PORTA
A2:         OUT   DX,AL
            INC   AL          ;next step
            CMP   AL,0        ;
            JZ    A1          ;if zero check for the key press
            MOV   CX,02FF     ;delay (for fast CPUs increase value in CX)
WT:         LOOP  WT          ;let DAC recover
            JMP   A2          ;create next step
STOP:       MOV   AH,4CH      ;go to DOS
            INT   21H
```

The following program sends the values to DAC continuously.

```
;in data segment
TABLE DB      128,192,238,255,238,192,128,64,17,0,17,64,128
;in code segment
            MOV   AL,80H            ;make all ports as output
            MOV   DX,CONPORT
            OUT   DX,AL
A1:         MOV   CX,12             ;count
            MOV   BX,OFFSET TABLE
            MOV   DX,PORTA          ;port A address
NEXT:       MOV   AL,[BX]
            OUT   DX,AL
            INC   BX
            CALL  DELAY             ;let DAC recover
            LOOP  NEXT
            JMP   A1                ;do it again
```

Notice that the above program does not check for the key press which means it is an infinite loop. If you run the above values on the circuit of Figure 1, you will see a very crude sine wave.

Activity 2

First generate (calculate) a table similar to Table 12-15 of the textbook for sine values of 4-degree increments. Then after making sure that your circuit setup in Activity 1 works, write and run a program in Assembly language to create the sine wave on the oscilloscope. Your program must check for key press. If any key is pressed, it will go back to DOS. Put a delay in between each outputting of values to DAC and observe the sine wave on the scope. Change the delay values and state your conclusions. Also in Figure 1, change the POT value and monitor the output sine wave on the scope.

Activity 3

Repeat Activity 2 using C language. In this program, we can use the "sin ( )" function (part of the "math.h" library) to calculate each value instead of using a look-up table. In your program, make the increment 1 degree instead of 4 degree. Compile and run the program in Example 12-15 in the textbook to verify its operation.

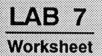
1. Define the following terminology in DAC.

   (a) resolution

   (b) full-scale voltage output

   (c) settling time

2. In Figure 1, find V_{out} for the following inputs.

   (a) 11001100

   (b) 10001111

3. To get a smaller step size, we need DAC with _____ (more, less) data bit inputs

4. In Figure 1, assume that R = 2.5 K ohms. Calculate V_{out} for the following binary inputs.

   (a) 11000010

   (b) 01000001

   (c) 00101100

   (d) 11111111

LAB 8

INTERFACING ANALOG-TO-DIGITAL CONVERTER (ADC) TO PC VIA EXPANSION SLOT

Objectives

» To interface an ADC to the 8255

References

Mazidi, Volumes I & II: Chapter 12
Data Acquisition Linear Devices Databook, National Semiconductor Corp.

Materials

80x86 IBM (or compatible) computer
MASM (or compatible) assembler
MicroSoft CodeView
PC Interface Trainer
ADC804
R = 10k ohms, C =150 pF

Activity 1

This activity tests the 804 connection. On the breadboard, set up the circuit shown in Figure 1. This setup is called free running test mode and is recommended by the manufacturer. In that figure we use a potentiometer to apply a 0-to-5 V analog voltage to the input V_{in} (+) of the 804 ADC and the binary outputs are monitored on the LEDs of the digital trainer. It must be noted that in free running test mode the CS input is grounded and the WR input is connected to the INTR output. However, according to National Semiconductor's databook "the WR and INTR node should be momentarily forced to low following a power-up cycle to guarantee operation". You must get this circuit working before you can go on to the next activity and many of the subsequent labs on the sensor connection.

Set the potentiometer at different points and record the digital binary output of the ADC indicated on the LEDs. In each case, use the voltammeter to measure the POT voltage and record it in the table in Question 1 of the worksheet. Examine your data to see how closely they match.

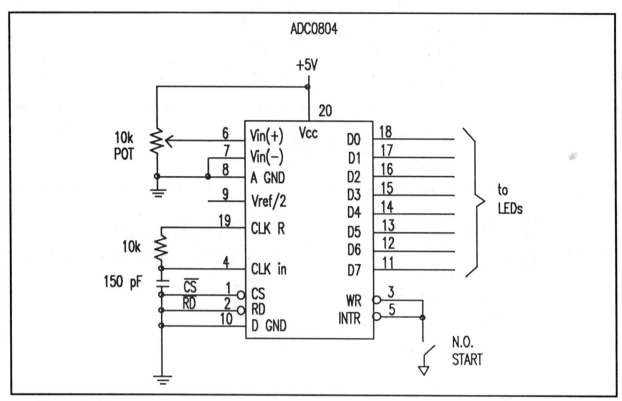

Figure 1. Testing 804 in Free Running Mode

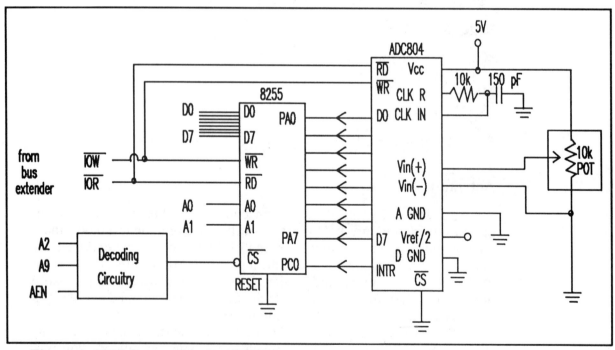

Figure 2. 8255 Connection to ADC804

Activity 2

After making sure that your ADC set up in Activity 1 works properly, connect the ADC804 to the 8255-to-PC interface as shown in Figure 2. In DEBUG, assemble and run the following program to see if the ADC connection to the 8255 works correctly.

```
-A 100
        MOV   AL,99H          ;ports A and C as input
        MOV   DX,303H         ;control port address
        OUT   DX,AL           ;initialize ports
A1:     MOV   DX,302          ;port C address
        IN    AL,DX           ;get intr status
        AND   AL,00000001     ;mask all except  PC0
        CMP   AL,00000001     ;is it end of conversion (or intr low)?
        JE    A1              ;if no keep checking PC0
        ;the conversion is finished, next get the data from port A
        MOV   DX,300H         ;port A address
        IN    AL,DX           ;8-bit binary data representing analog input
        INT 3                 ;stop, now AL has the analog input
-G=100 {return}
```

Now change the position of the potentiometer and run the above program and examine the contents of the AL register. Repeat this process several times to see various digital outputs generated by the 804 ADC chip.

Activity 3

After making sure that Activity 2 works properly, write a program (in C or Assembly) to get the analog input and display the result on the PC screen. Notice that the ADC output is between 0 to FFH where FFH is for a full-scale input of 5 volts. However, this must be converted to decimal and then to ASCII in order to be displayed on the PC screen. As you change the potentiometers, the output should change, indicating the value of the analog input. Your program must check the key press and if Esc is pressed, it goes back to DOS. Using C language you can convert the analog input value to millivolts (or even volts). However, such conversion in Assembly language is tedious.

Note: Do not disassemble this circuit. It is used in many of the subsequent labs on interfacing sensors to the PC.

Name: _____
Date: _____
Class: _____

1. Fill in the following table for Activity 1. The table shows V_{out} (in binary) for various V_{in} on the 804 as well as the LED output.

Case	V_{in} Applied to V_{in} (+) (measured from POT)	D7	D6	D5	D4	D3	D2	D1	D0	V_{out} (#steps x step size)
1										
2										
3										
4										

2. Indicate the direction of pins WR, RD, and INTR from the point of view of the ADC804 chip. What is the direction from the point of view of the microprocessor?

3. Give the three steps for converting data and getting the data out of the 804. State the status of the CS, RD, INTR, and WR pins in each step.

4. Assume that $V_{ref}/2$ is connected to 1.28 V. Find the following.
   (a) step size
   (b) maximum range for V_{in}
   (c) D7 - D0 values if V_{in} = 1.2 V
   (d) V_{in} if D7 - D0 = 11111111
   (e) V_{in} if D7 - D0 = 10011100

5. Assume that $V_{ref}/2$ is connected to 1.9 V. Find the following.
   (a) step size.
   (b) maximum range for V_{in}
   (c) D7 - D0 values if Vin = 2.7 V
   (d) Vin if D7 - D0 = 11111111
   (e) V_{in} if D7 - D0 = 11011101

6. In Activity 1, often the V_{in} measured across the potentiometer is not exactly the same as the value calculated from the product of the step size and number of steps as indicated by the LEDs. Give the reason for this discrepancy.

7. In question 6, give two ways to make the values as close as possible if you are designing a circuit.

LAB 9

INTERFACING A TEMPERATURE SENSOR TO A PC

Objectives

» To interface the LM35 temperature sensor to a PC via the 8255

References

Mazidi, Volumes I & II: Chapters 12 and 28
Data Acquisition Linear Devices Databook, National Semiconductor Corp.

Materials

80x86 IBM (or compatible) computer
MASM (or compatible) assembler
MicroSoft CodeView
PC Interface Trainer
LM35 (or LM34)
ADC804
LM336-2.5
10K POT
1K, 1.5K, and 10K resistors

Activity 1

Connect the LM35 or (LM34) of your choice to V_{CC} and ground. Then use a voltmeter to measure the voltage at V_{out}. Heat up or cool down the sensor to examine the changes on the digital read-out of the voltmeter.

Activity 2

To connect LM35 of Activity 1 to the ADC804, set up the circuit in Activity 1 of Lab 8 and replace the potentiometer with the LM35 (or LM34). Then heat up or cool down the sensor and examine the LEDs. The LEDs should change as you change the temperature.

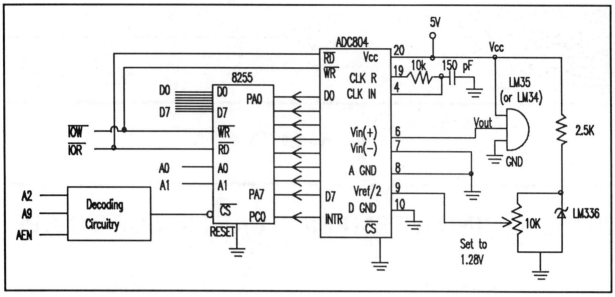

Figure 1. Temperature Sensor Connection to ADC804

Activity 3

Set up the circuit shown in Figure 1 and write a program (in C or Assembly) to display the temperature on the PC monitor continuously. Your program should check for key press and if Esc is pressed, it goes back to DOS. (You can modify the program and the circuit from Activity 3 of Lab 8).

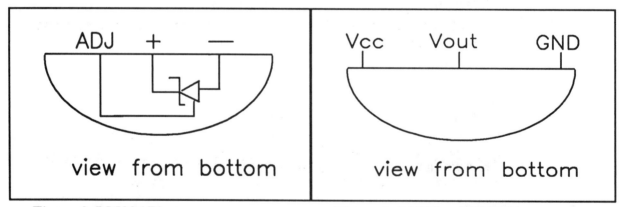

Figure 2. LM336 Pins   **Figure 3. LM34 or LM35 Pins**

Note: In Figures 2 and 3, "view from bottom" means that the pins are pointing at you.

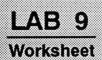

LAB 9
Worksheet

Name: _____
Date: _____
Class: _____

1. What is a transducer?

2. What is the form of the output of a transducer?

3. The preprocessing of transducer signals to be fed into an ADC is called _____.

4. The LM35 and LM34 produce _____ mV output for every degree change in temperature.

5. The LM35/LM34 is a _____(linear, nonlinear) device. What is the advantage of a linear device?

6. Explain signal conditioning and its role in data acquisition.

LAB 10

PROGRAMMING THE 8253 TIMER OF THE PC INTERFACE TRAINER

Objectives

» To test the 8253 timer of the PC Interface Trainer
» To play music using the 8253 timer

Reference

Mazidi, Volumes I & II: Chapter 13

Materials

PC Interface Trainer
scope
LM386 from National Semiconductor
10K Potentiameter
10 uF capacitor (polarized)
10K ohms potentiometer
10 ohms resistor
210 uF polarized capacitor
0.05uF capacitor
speaker

In this lab, we first test the 8253/54 timer installed on the PC Interface Trainer to make sure it is working. Then we connect one of the counters of the 8253 to the speaker and play music.

Examine the 8253/4 timer on the Trainer. It uses port addresses 304H to 307H. The Gate, CLK, and OUT signals for all three counters of the 8253/54 are available on both headers H4 and H8. One is a SIP socket and the other one is a male connector. We also have a TTL-output crystal oscillator. The square wave TTL frequency output from this oscillator is also available on H4 and H9. They are labeled as OSC. The frequency of the OSC is either 1 MHz or 2 MHz. Find which one is installed on yours. Use a scope to verify the frequency of this oscillator. This must be done before you can perform any of the following activities.

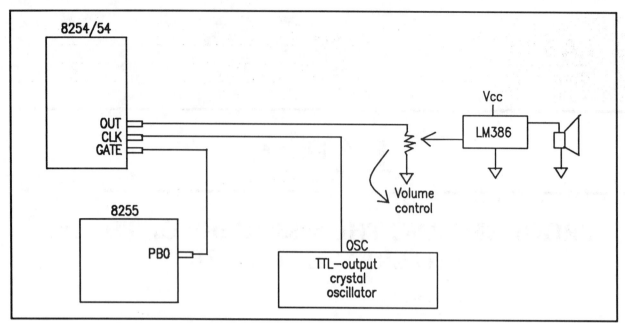

Figure 1. 8253/54 Connection to Audio Amplifier and Speaker

Activity 1

To test the operation of the 8253/54 timer, connect the OSC pin to the CLK pin of counter 0 (or counter 1 or counter 2, whichever you want). Then connect the GATE pin of the counter to V_{CC} to enable the counter. Notice that you have V_{CC} SIP socket pins on the Trainer itself.

Now write a program to divide the CLK frequency by a number (1500, for example). Use the scope to see and compare the frequency on the CLK and OUT pins to verify it is divided. The following could be a skeleton of such a program in DEBUG. This will divide the CLK0 frequency by 1500 (1500 which is 05DC in Hex) in mode 3. *Reminder:* In DEBUG the numbers are in hex.

```
MOV AL,36
MOV DX,307
OUT DX,AL
MOV AL,DC
MOV DX,304
OUT DX,AL
MOV AL,05
NOP
NOP
OUT DX,AL
INT 3
```

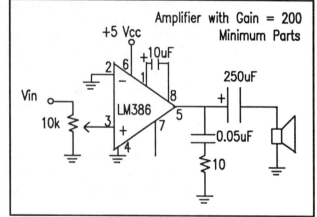

Figure 2. Amplifier

Activity 2

After making sure the 8253/54 timer works in Activity 1, connect OUT of the 8253/54 to the LM386 audio amplifier and then to a speaker, as shown in Figure 2. Also connect the Gate of the counter to PB0 (Port B of the 8255). Write a program to play the music of your choice. Use the Port B bit to disable the 8253 after the music has finished playing.

1. For the 8253/54 of the PC Interface Trainer, give the port addresses used for each of the following:

   Port Address (in Hex)

   Counter 0
   Counter 1
   Counter 2
   Control reg.

2. Which headers provide the signals to the counters of the 8253/54?

3. In the above question, what is the purpose of the male header?

4. Which header provides access to the output of the oscillator?

5. Modify the program of Activity 1 (in DEBUG), for counter 2. Use the BCD option in the control word.

6. In Activity 2, why do we connect the Gate to PB?

7. Cascade two of the counters of the 8253/54 timer to divide the clock by 100,000. Write the program and verify it.

SECTION 5

Serial and Parallel Port Interfacing

This section explores interfacing devices to the PC via the serial and parallel ports. In the first two labs we explore serial communication between two PCs. Then we present a lab which demonstrates how to access a PC's parallel port data bus. After that, devices such as LCDs, DAC, and stepper motors are interfaced to a PC via the parallel port.

LAB 1

PC-TO-PC SERIAL DATA COMMUNICATION

Objectives

» To connect two PCs together through their COM ports and transfer data between them using a Windows terminal program
» To identify all the COM ports of the PC and the I/O base address for each one using DEBUG
» To write a C program to find and display the base I/O port address for each COM port installed

References

Mazidi, Volumes I & II: Chapter 17, Section 28.2, Appendix A

Materials

Two 80x86 IBM PC (or compatible) computers, each with a free serial port. A break-out box, or null modem adapter from Radio Shack (part no. 26-1496 for DB25 male/DB25 female, or 26-264 for DB9 male/DB9 female, you may also need a gender changer)

Activity 1

In this activity we connect the COM ports of two PCs together and use Window's terminal program to transfer data between them. The following steps need to be taken.

1. Examine the back of both PCs to locate the serial port (DB-25 or DB-9) male socket. Make sure you identify the COM number such as COM1, COM2, and so on. Use the MSD utility (it comes with DOS) to identify the COM port number and its use. Normally, COM1 is used for the mouse and COM2 is available. In that case, we can use COM2.

Note: Use a break-out box for steps 2 and 3 if you have access to one, or use a null modem adapter from Radio Shack.

2. Connect three signals between the two COM ports as follows. See Figure 1.
   (a) TXD of PC#1 to RXD of PC#2
   (b) TXD of PC#2 to RXD of PC#1
   (c) Signal Ground of PC#1 to Signal Ground of PC#2
3. (a) For PC#1, jumper a wire from the RTS pin to CTS pin. Also jumper a wire from DSR pin to DTR pin. See Figure 1.
   (b) Repeat step 3a for PC#2

4. Using File Manager of Windows, go to the Windows directory, locate the "terminal.exe" program, and run it.
5. Click on "communications" for settings in the terminal program. Examine your choices.
6. Choose 9600 baud rate, 8 data bits, 1 stop bit, none for parity, none for flow control, and COM2.
7. Now press a key on one PC and see if it is transferred to the other one.

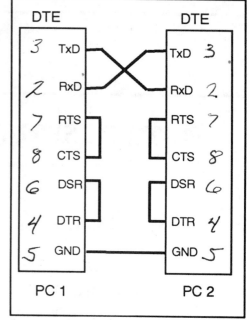

Figure 1. DTE-DTE Connection

Do not disconnect the PCs. This connection will be used in the next lab.

Activity 2

Use the dump command of DEBUG to dump the BIOS data area 40:0000 - 0007 and get the base I/O port address of the PC's installed COM port. Complete the table shown in Question 6. In the table, use the MSD utility (it comes DOS 5 and higher) to verify the base I/O address and get the answer for usage.

Activity 3

Study Section 28.2 of the textbook and write a C program to list the base I/O address of each of the COM ports in your PC. If a COM port is not installed, it should indicate that.

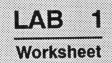

LAB 1

Worksheet

Name: _____

Date: _____

Class: _____

1. How do we distinguish between the parallel port (LPT) and serial port (COM) connectors on the back of the PC if both use 25-pin connectors?

2. List three main signal lines needed to transfer and receive data in the RS232 serial port. List the abbreviation for each signal, what it stands for, and its function.

3. List all the signal lines associated with receiving data in the RS232 serial connector. List the abbreviation for each signal, what it stands for and its function.

4. Why do you think IBM introduced a 9-pin connector for serial ports in the PC?

5. The 9-pin connector for the COM port on the back of the PC is _____ (male,female). How about the 25-pin COM port?

6. Fill in the following table for one of the PCs used in this experiment.

	Your PC		
	Installed? (Y/N)	Base I/O Address	Connector Type
COM1			
COM2			
COM3			
COM4			

LAB 2

MORE PC-TO-PC SERIAL DATA COMMUNICATION

Objectives

» To code a program in Assembly language making two PCs communicate back and forth where each sends and receives data
» To code a program in C making two PCs communicate back and forth where each sends and receives data

References

Mazidi, Volumes I & II: Chapters 4, 17 and 28

Materials

Two 80x86 IBM PC (or compatible) computers, each with a free serial port
MASM (or compatible) assembler
Serial cable or break-out box
Turbo C Compiler

Activity 1

Part (a)

First make sure that your PC-to-PC COM port connection in Activity 1 of the previous lab works properly. Then use an assembler such as MASM or TASM to write and run an Assembly language program to make two PCs talk back and forth. This program is listed in Chapter 17.

Part (b)

Modify Part (a) to insert a line feed upon detecting a carriage return.

Activity 2

You can make Activity 1 even more interesting by making each PC have two screen windows: one for the data received and the other one for data being picked up from its own keyboard and sent to the other PC.

1. Discuss the difference between half duplex and full duplex.

2. In this lab, the data transfer is done using _____ (full, half) duplex.

3. In order to perform full duplex, we must have how many data lines, and why?

4. Show the flow chart of the program in Activity 1.

LAB 3

ACCESSING A PC'S PARALLEL PORT DATA BUS

Objectives

» To use the data pins of a PC's parallel port as an output port
» To use the control pins of a PC's parallel port as an output port

References

Mazidi, Volumes I & II: Chapters 18 and 28
Jan Axelson, "How to Use a PC's Parallel Port," *MicroComputer Journal*, May/June 1994.
www.lvr.com

Materials

80x86 IBM (or compatible) computer
MASM (or compatible) assembler
MicroSoft CodeView
Digital trainer
74LS244

In this lab we examine the data pins and control pins of the printer's parallel port and then show how to access them. Every PC has a parallel port accessible through the DB25 connector. The printer cable end going to the PC is DB25 male while the end going to the printer is a centronics connector. The pin-out for both are shown in Chapter 18 of the textbook.

Unidirectional data bus of the printer port

In every 80x86 IBM PC/compatible computer made since 1981, the parallel port supports a unidirectional data bus. This means that from the PC, data can be sent out, but not brought into the PC. However, starting with the PS/2 models (introduced in 1987), IBM made the data pins of the parallel port a bidirectional bus meaning that data can be sent out or brought into the PC. Note that not all compatible PCs made since 1987 support a bidirectional parallel data bus. For this reason, you should not try bringing data into a PC via the parallel port's data pins unless you are sure that the data bus is bidirectional. To do so can cause damage to your parallel port. This is also the reason that in this lab we use the data pins of the parallel port only for data output. See a series of excellent articles by Jan Axelson in *MicroComputer Journal* (formerly ComputerCraft) starting in May/June 1994.

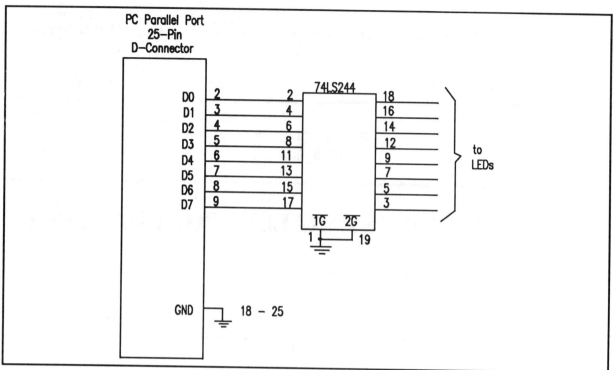

Figure 1. Accessing Parallel Port Data Bits

Buffering the printer's port data bus

To avoid damaging the data bus of the parallel port, the data buses are connected to a 74LS244. This not only isolates the data pins but also boosts the driving current of the data bus. Next we see how to access the data bus portion of the parallel port.

Activity 1

As discussed in Chapter 18 of the textbook, three I/O port addresses are assigned to each parallel port: one for the data signals, one for the status signals, and one for the control signals. In many PCs, LPT1 has the base I/O address of 03BCH; therefore, port 3BCH is for data, 3BDH is for status, and 3BEH for control signals. In some PCs, the base I/O address for LPT1 is 0378H and for this reason the base I/O address belonging to a given LPT must be fetched from the BIOS data area.

The following steps show how to access the data section of a parallel port. For this activity, go through these steps and make sure they work, since they are used in subsequent labs in this section.

1. On the breadboard, set up the circuit shown in Figure 1. Notice that the data pins of the PC parallel port are buffered first by the 74LS244 before it is connected to the LEDs of the digital trainer. Make sure that all the grounds are connected to the same point. Notice that the PC parallel port has 8 ground pins (pins 18 - 25 on the DB25) and that they are connected to the ground on the breadboard.

2. In DEBUG, dump the BIOS data area belonging to this parallel port to get the base address of the parallel port. You can also get it by using the MSD software which comes with DOS.

3. In DEBUG, assemble and run the following program to send data to the LEDs (Recall that in DEBUG, H for hex need not be typed. Also, do not type

comments, they are given for clarity). After typing in the program below, set
G=100 and run it. Then examine the LEDs; you should see 0101 0101 (55
hex).

```
-A 100
MOV   DX,3BCH      ;base i/o address for LPT (put the result of step 2)
MOV   AL,55H       ;AL=55H
OUT   DX,AL        ;send it to data pins of LPT port
INT   3            ;stop and dump the registers
```

4. In the line "MOV AL,55H" change the data 55H to AAH and run the program
again. This step ensures that the data pins work for both low and high and
no bit is stuck to a 0 or 1 state.

Activity 2

After making sure that your circuit in Activity 1 works properly, write
a program (in C or Assembly) that counts in binary and sends each count to
the LEDs. This is an 8-bit counter and the maximum number is 1111 1111
(FF in hex or 255 decimal).

Provide a sufficient time delay in between each count so that you can
see the count on the LEDs. In a C program you can use the delay() function
(see Chapter 28 of the textbook).

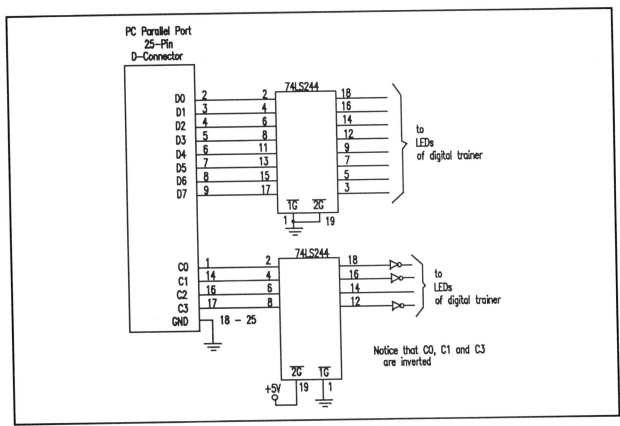

Figure 2. Accessing Parallel Port Data and Control Bits

Accessing control bits of the PC's parallel port

In the first two activities, we accessed only the data pins of the PC's
parallel port. We can also use the control pins of the parallel port for the
purpose of outputting data. We first must make sure that they are buffered
with the 74LS244. This is shown in Figure 2.

Activity 3

After buffering C0 - C3 of the control port, connect them to the LEDs as shown in Figure 3. Notice in Figure 2 that we are using only the lower four bits of the control port: C0, C1, C2, and C3. Assemble and run the following program (in DEBUG) to test the connection.

```
-A 100
MOV   DX,3BEH      ;LPT'S control address (use the address for your PC)
MOV   AL,5         ;AL=5
OUT   DX,AL        ;send it to control pins of the LPT port
INT   3            ;stop and dump the registers
```

Run the above program and examine the LEDs. Then set AL=00001010 (0A in hex) and run the program again. Now you are ready to combine this activity with Activity 2 to make a 12-bit counter.

Activity 4

Using the 8 bits of the data port (from Activity 2) and the 4 bits of the control port (from Activity 3), write and run a program (in C or Assembly) that counts in binary and sends each count to 12 LEDs. This is a 12-bit counter and the maximum number is 1111 1111 1111 binary (FFF in hex or 4097 decimal). While the lower 8 bits go to the data pins, the upper 4 bits go the control pins of the LPT port.

Notes from the Trenches

When accessing the PC's parallel port for data acquisition, your program should get the base I/O address from the BIOS data area. This makes the program dynamic and able to be run on any PC. In Assembly language, we can use the following code:

```
PUSH  DS            ;save DS
PUSH  AX            ;save AX
SUB   AX,AX         ;AX=0
MOV   DS,AX         ;DS=0 for BIOS data area
MOV   DX,[408]      ;get the LPT1 I/O base address
POP   AX            ;restore AX
POP   DS            ;restore DX
;now DX has the I/O base address of LPT1
```

In C we can use the following code:

```
main()
{
        ...
        unsigned int far *xptr;
        xptr = (unsigned int far *) 0x00000408;
        outp(*xptr,mybyte);    /* send mybyte to LPT's data port */
        ...
}
```

For more on this subject, see Chapter 28 of the textbook.

1. In a given PC, the I/O base address for one of the printer's parallel ports is 278H. Indicate the port addresses for each of the following.
   (a) data pins
   (b) status pin
   (c) control pins

2. True or false. The parallel port on every 80x86 IBM PC/compatible supports data outputting using its data pins.

3. True or false. The parallel port on every 80x86 IBM PC/compatible support data outputting and data inputting using its data pins.

4. True or false. From the point of view of the PC, the control pins of the parallel port in all PCs are outputs.

5. Using DEBUG, show how you find the base I/O address for LPT1, LPT2, and LPT3. Which one did you use in this lab?

LAB 4

INTERFACING AN LCD TO THE PC VIA A PARALLEL PORT

Objective

» To interface an LCD to the PC's parallel port

References

Mazidi, Volumes I & II: Chapter 18
Lab 4 in Section 4 of this lab manual
Lab 1 in Section 5 of this lab manual

Materials

80x86 IBM (or compatible) computer
MASM (or compatible) assembler
MicroSoft CodeView
All the parts in Lab 4 of Section 4 and Lab 1 of Section 5

In Lab 4 of Section 4, we discussed the LCD in detail and showed how it can be interfaced with a PC via the expansion slot. Make sure that you have studied Lab 4 before embarking on this lab. In Lab 1 of this section we showed how to use the data pins and control pins for outputting data. In this lab we build on our experiences from those two labs and interface an LCD to a PC's parallel port.

Activity 1

Breadboard the circuit in Figure 1, connecting the LCD to the PC's parallel port. This is after you have made sure the circuit in Activity 4 of Lab 1 of this section works properly. Then write a program to display your last name and first name on the LCD.

Note: In this lab we use the data pins of the parallel port only for outputting. Therefore you cannot use the data pins for the purpose of inputting the busy flag of the LCD. For this reason you must send code and data to the LCD's data pins without checking the busy flag. Although this is not recommended by the LCD manufacturer it will work if sufficient time delays are provided before the next data is sent to the LCD. This is discussed in Lab 4 of Section 4 in this lab manual.

Activity 2

After successfully completing Activity 1, pick one of Activities 3 or 4 of Lab 4 in Section 4 and modify it for the parallel port connection to LCD.

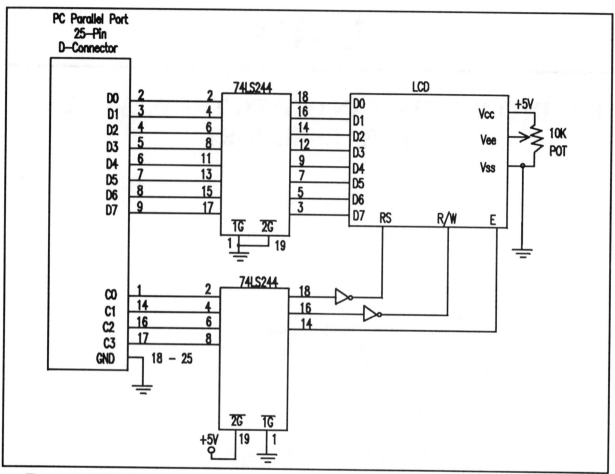

Figure 1. LCD Connection to Parallel Port

LAB 5

INTERFACING DAC TO PC VIA PARALLEL PORT

Objectives

» To interface a digital-to-analog converter to a PC via a parallel port
» To draw square or lisajous figures or the shape of an object such as a car on an oscilloscope using DAC

References

Mazidi, Volumes I & II: Chapter 18
Lab 7 of Section 4 of this lab manual

Materials

80x86 IBM (or compatible) computer
MASM (or compatible) assembler
MicroSoft CodeView (or Turbo debugger)
All the parts in Lab 7 of Section 4 in this lab manual
Two-channel oscilloscope

In Lab 1 of this section, we showed how to use the data pins of the parallel port to send out data. We also discussed DAC interfacing via the expansion slot in Lab 7 of Section 4. In this lab, we build on the experiences of those two labs in interfacing a DAC to the PC's parallel port.

Activity 1

Breadboard the circuit in Figure 1. This the same as Figure 1 in Lab 7 of Section 4, except that it is connected to the data portion of a parallel port. Modify the program in Activity 2 of Lab 7 (Section 4) to generate a sine wave and display it on an oscilloscope.

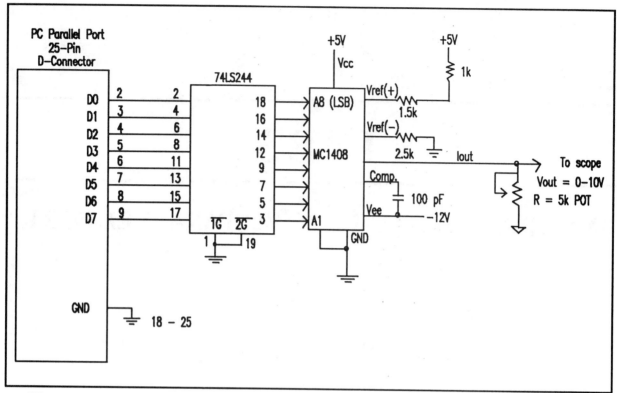

Figure 1. Parallel Port Connection to DAC

Activity 2

Complete the following steps to generate a two-dimensional drawing on the oscilloscope. Your oscilloscope must have two channels.

1. Connect two DACs to the PC, one via the PC's parallel port and the other one via the expansion slot through the PC Interface Trainer as was used in Lab 7 of Section 4.
2. Connect the output of DAC number 1 to the X input of the oscilloscope.
3. Connect the output of DAC number 2 to the Y input of the oscilloscope.
4. Set your scope for the X-Y option.
5. On each DAC, generate the sine wave. By examining the scope you will see what are called lisajous figures. Lisajous figures are created as a result of two perpendicular harmonic signals displayed on the scope at the same time. If you are not familiar with lisajous figures, connect two signal generators to the X and Y inputs of the scope and send two sine waves, one to each input. Change the magnitude and frequency of the sine waves and examine the lisajous figures on the scope. You will see some very interesting dancing curve signals.

Activity 3 (optional)

Instead of generating lisajous figures, you can draw a square or any shape such as a car or space shuttle on the scope. You need to make a table where the values for the X and Y axis are stored. In such a table, each point of the figure is represented by an x and y value where the x value is followed by the y value. The x86 can be programmed to fetch each byte and send it to each DAC as shown next.

```
A1:     MOV   SI,OFFSET TABLE      ;load the table address
        MOV   AH,1                 ;check for keypress
        INT   16H                  ;using INT 16
        JNZ   EXT                  ;exit to DOS if a key is pressed
A2:     MOV   AL,[SI]              ;get the x value
        CMP   AL,0                 ;is it end of table?
        JZ    A1                   ;if yes then start all over
        MOV   DX,MYPORT_1          ;otherwise send it to DAC
        OUT   DX,AL
        CALL  DELAY                ;let DAC recover
        INC   SI
        MOV   AL,[SI]              ;get the y value
        MOV   DX,MYPORT_2
        OUT   DX,AL
        INC   SI
        CALL  DELAY                ;let DACs recover
        JMP   A2                   ;get the next x and y
EXT:
```

Notice in the above program that we check for 0, the end of the table, to start all over again. Therefore, make sure that your table has 0 (ASCIIZ) as the last element. Notice also that to draw any figure on the scope we must send x and y values continuously. To avoid an infinite loop, we make sure that when a key is pressed the program exits to DOS.

In the case of generating a square on the scope, you do not need a table for the values. Instead, set the initial x and y values and increment y to draw a vertical line. Similarly, for the horizontal line, increment x.

The above activity gives you an understanding of how graphics are generated on a PC monitor. See Chapter 26 of the textbook for examples of how to use INT 1OH to generate graphics.

LAB 6

STEPPER MOTOR INTERFACING VIA THE PARALLEL PORT

Objective

» To interface a stepper motor to the PC via a parallel port

References

Mazidi, Volumes I & II: Chapters 12 and 18
Lab 5 of Section 4 of this lab manual

Materials

80x86 IBM PC (or compatible) computer
MASM (or compatible) assembler
MicroSoft CodeView
All the parts of Lab 5 of Section 4

Before you can embark on this lab, familiarize yourself with Lab 5 of Section 4.

Activity 1

In Lab 5 (Section 4) of this lab manual, we discussed details of stepper motor interfacing. In Lab 1 of this section, we discussed how to use the parallel port to send data out. Connect the stepper motor to the data portion of the PC's parallel port as shown in Figure 1. Again, you must buffer the data pins of the parallel port with the 74LS244. First, write a simple program (similar to Activity 1 in Lab 5 of Section 4) to control the stepper motor from the parallel port. After making sure that it works, modify the programs for Activity 3 (or 4) and run it to control the stepper motor from the PC's parallel port.

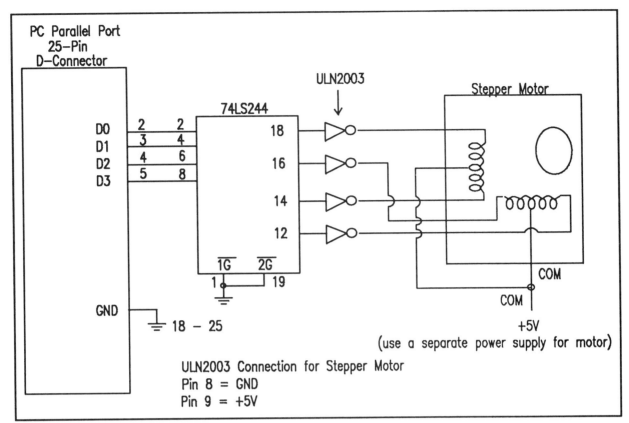

Figure 1. Stepper Motor Connection to Parallel Port

Appendix A

FLOWCHARTS AND PSEUDOCODE

Flowcharts

If you have taken any previous programming courses, you are probably familiar with flowcharting. Flowcharts use graphic symbols to represent different types of program operations. These symbols are connected together into a flowchart to show the flow of execution of the program. Figure 1 shows some of the more commonly used symbols. Flowchart templates are available to help you draw the symbols quickly and neatly.

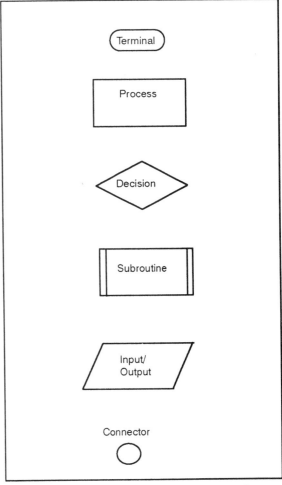

Figure 1: Commonly Used Flowchart Symbols

Pseudocode

Flowcharting has been standard practice in industry for decades. However, some find limitations in using flowcharts, such as the fact that you can't write much in the little boxes, and it is hard to get the "big picture" of what the program does without getting bogged down in the details. An alternative to using flowcharts is pseudocode, which involves writing brief descriptions of the flow of the code. Figures 2 through 6 show flowcharts and pseudocode for commonly used control structures. Then we show a few examples of flowcharts and pseudocode for three programs in the textbook.

Structured programming uses three basic types of program control structures: sequence, control, and iteration. Sequence is simply executing instructions one after another. Figure 2 shows how sequence can be represented in pseudocode and flowcharts.

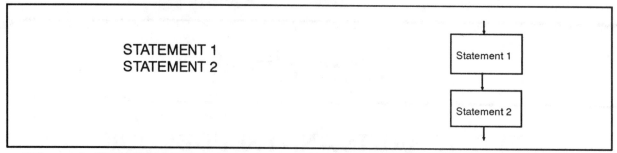

Figure 2: SEQUENCE Pseudocode versus Flowchart

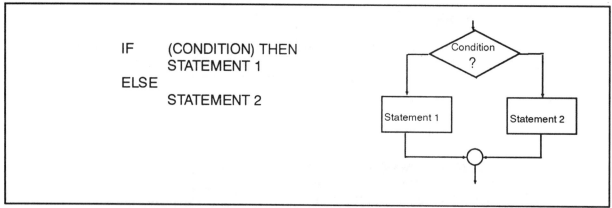

Figure 3: IF THEN ELSE Psuedocode versus Flowchart

Figures 3 and 4 show two control programming structures: IF-THEN structures and IF-THEN-ELSE structures in both pseudocode and flowcharts.

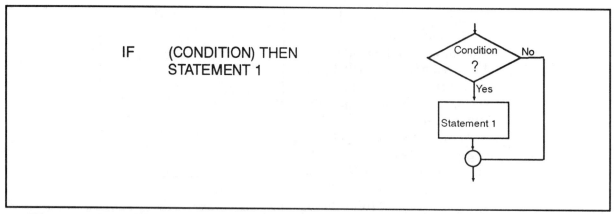

Figure 4: IF THEN Psuedocode versus Flowchart

Note in Figures 2 through 6 that "statement" can indicate one statement or a group of statements.

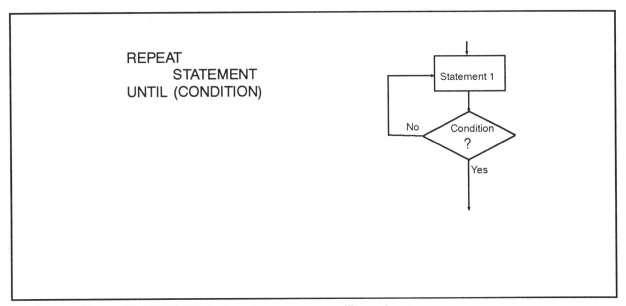

Figure 5: REPEAT UNTIL Pseudocode versus Flowchart

Figures 5 and 6 show two iteration control structures: REPEAT UNTIL and WHILE DO. Both structures execute a statement or group of statements repeatedly. The difference between them is that the REPEAT UNTIL structure always executes the statement(s) at least once, and checks the condition after each iteration, whereas the WHILE DO may not execute the statement(s) at all since the condition is checked at the beginning of each iteration.

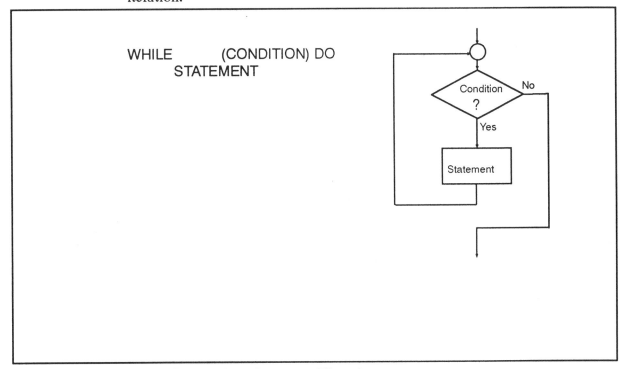

Figure 6: WHILE DO Pseudocode versus Flowchart

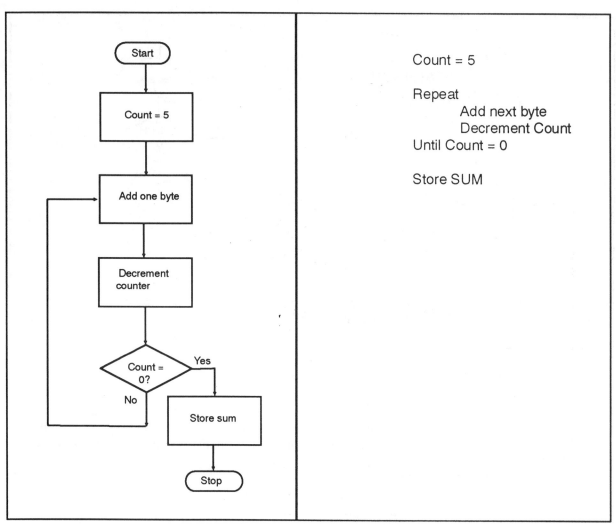

Flowchart for Program 2-1   **Pseudocode for Program 2-1**

This example shows the flowchart versus the pseudocode for Program 2-1 of the textbook. In this example, more program details are given than one usually finds. For example, this shows steps for initializing and decrementing counters. Another programmer may not include these steps in the flowchart or pseudocode. Notice that housekeeping chores such as initializing the data segment register in the MAIN procedure are not included in the flowchart or pseudocode.

It is important to remember that the purpose of flowcharts or pseudocode is to show the flow of the program and what the program does, not the specific Assembly language instructions that accomplish the program's objectives.

Notice also that the pseudocode gave the same information in a much more compact form than the flowchart. It is important to note that sometimes pseudocode is written in layers, so that the outer level or layer shows the flow of the program and subsequent levels show more details of how the program accomplishes its assigned tasks.

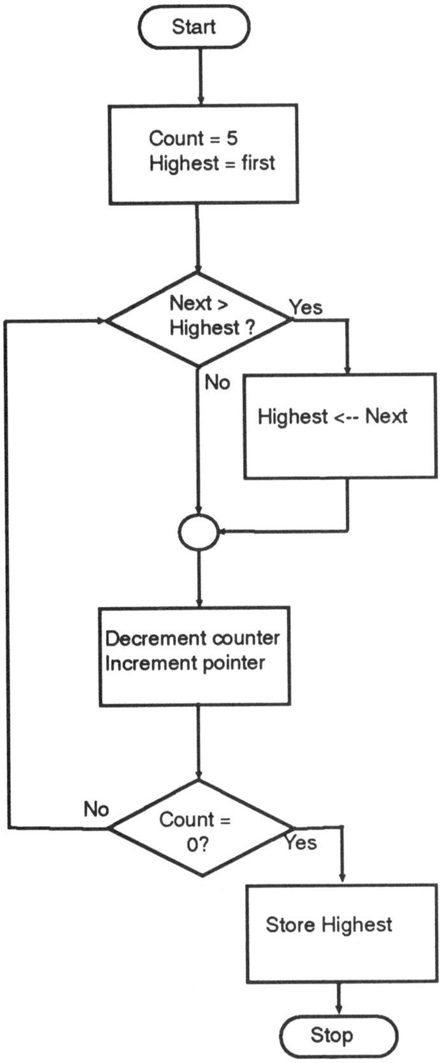


```
Count <-- 5
Highest <-- first item

Repeat
        IF Next > Highest
  THEN Highest <-- Next
        Decrement Count
Until Count = 0

Store Highest
```

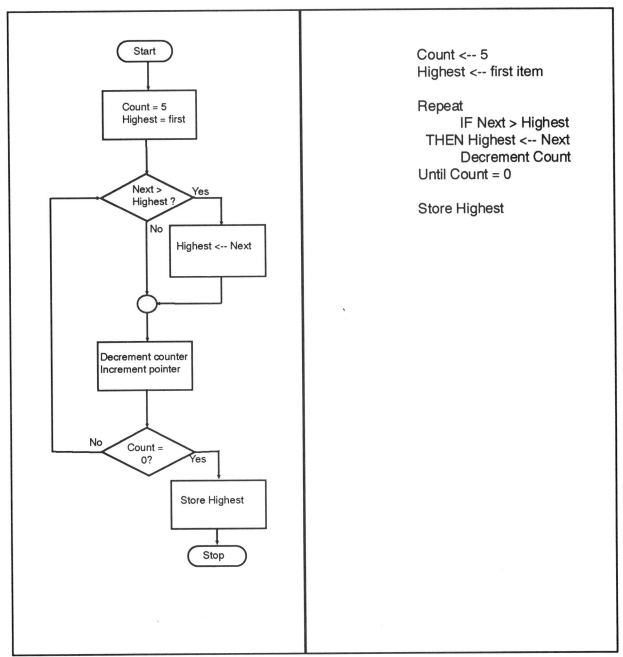

Flowchart for Program 3-3

Pseudocode for Program 3-3

This page shows the flowchart versus the pseudocode for Program 3-3 of the textbook.

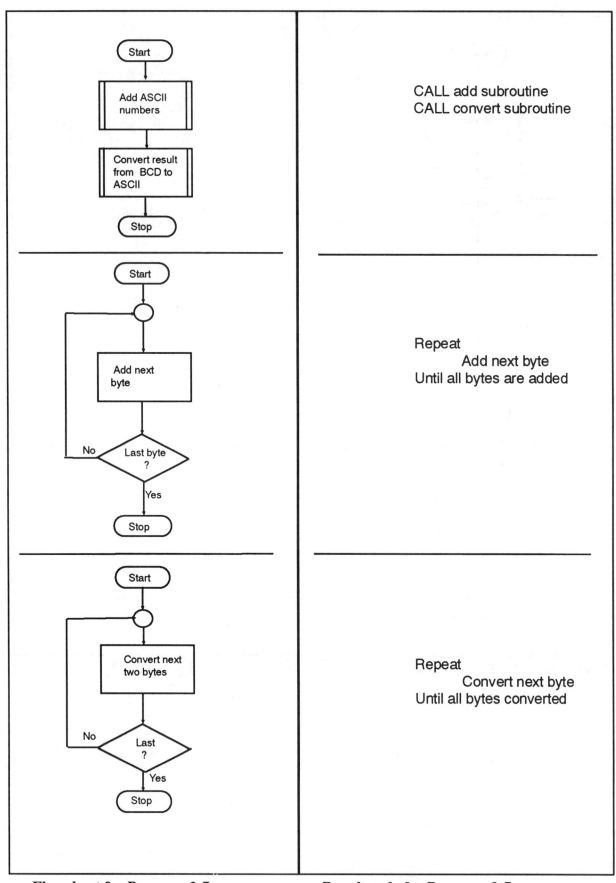

Flowchart for Program 3-7

Pseudocode for Program 3-7

CALL add subroutine
CALL convert subroutine

Repeat
      Add next byte
Until all bytes are added

Repeat
      Convert next byte
Until all bytes converted

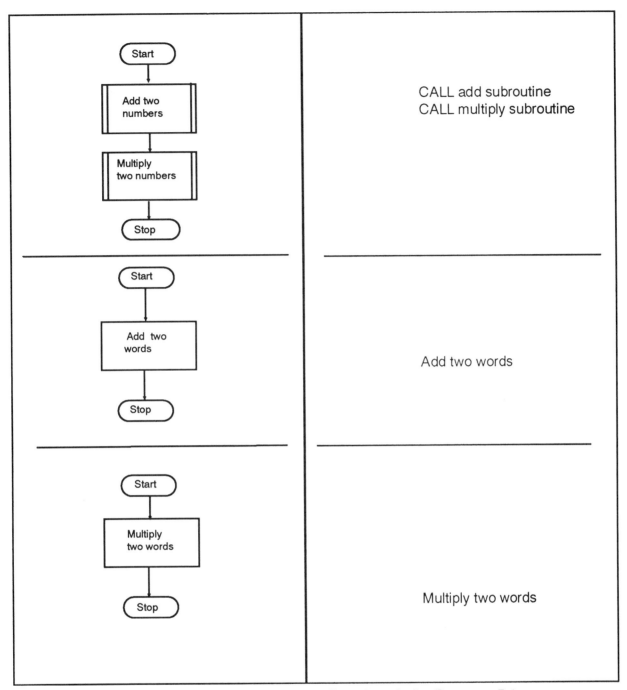

| Flowchart for Program 7-1 | Pseudocode for Program 7-1 |

Appendix B

THE BASICS OF WIRE-WRAPPING

Note: For this tutorial appendix, you will need the following:

Wire-wrapping tool (Radio Shack part number 276-1570).
30-gauge (30-AWG) wire for wire wrapping

The following describes the basics of wire wrapping.

1. There are several different types of wire-wrap tools available. The best one is available from Radio Shack for less than $10. The part number for Radio Shack is 276-1570. This tool combines the wrap and unwrap functions in the same end of the tool and includes a separate stripper. We found this to be much easier to use than the tools that combined all these features on one two-ended shaft. There are also wire-wrap guns, which are of course more expensive.

2. Wire-wrapping wire is available prestripped in various lengths or in bulk on a spool. The prestripped wire is usually more expensive and you are restricted to the different wire lengths you can afford to buy. Bulk wire can be cut to any length you wish which allows each wire to be custom fit.

3. There are a few different types of wire-wrap boards available. These are usually called perfboards or wire-wrap boards. These types of boards are sold at many electronics stores (such as Radio Shack). The best type of board has plating around the holes on the bottom of the board. These boards are better because the sockets and pins can be soldered to the board which makes the circuit more mechanically stable.

4. Choose a board that is large enough to accommodate all the parts in your design with room to spare so that the wiring does not become too cluttered. If you wish to expand your project in the future, you should be sure to include enough room on the original board for the complete circuit. Also, if possible, the layout of the IC on the board needs be done such that signals go from left to right just like the schematics.

5. To make the wiring easier and to keep pressure off the pins, install one standoff on each corner of the board. You may also wish to put standoffs on the top of the board to add stability when the board is on its back.

6. For power hook-up, use some type of standard binding post. Solder a few single wire-wrap pins to each power post to make circuit connections (to at least one pin for each IC in the circuit).

7. To further reduce problems with power, each IC must have its own connection to the main power of the board. If your perfboard does not have built-in power buses, *run a separate power and ground wire from each IC to the main power*. In other words, DO NOT daisy chain (chip-to-chip connection is called daisy chain) power connections, as each connection down the line will have more wire and more resistance to get power through. However, daisy chaining is acceptable for other connections such as data, address, and control buses.

8. You must use wire wrap sockets. These sockets have long *square* pins. The pins must be square so that the edges will cut into the wire as it is wrapped around the pin.

9. Wire wrapping *will not work on round legs*. If you need to wrap to components, such as capacitors that have round legs, you must also solder these connections. The best way to connect single components is to install individual wire-wrap pins into the board and then solder the components to the pins. An alternate method is to use an empty IC socket to hold small components such as resistors and wrap to the socket.

10. The wire should be stripped about 1 inch. This will allow for 7 to 10 turns for each connection. The first turn or turn-and-a-half should be insulated. This prevents stripped wire from coming in contact with other pins. This can be accomplished by inserting the wire as far as it will go into the tool before making the connection.

11. Try to keep wire lengths to a minimum. This prevents the circuit from looking like a bird nest. Be neat and use color coding as much as possible. Use only red wires for VCC and black wires for ground connections. Also use different colors for data, address, and control signal connections. These suggestions will make trouble shooting much easier.

12. It is standard practice to connect all power lines first and check them for continuity. This will eliminate one source of trouble later on.

13. It's also a good idea to mark the pin orientation on the bottom of the board. There are plastic templates available with pin numbers preprinted on them specifically for this purpose or you can make your own from paper. Forgetting to reverse pin order when looking at the bottom of the board is a very common problem when wire wrapping circuits.

14. To prevent damage to your circuit, place a diode (such as IN5338) in *reverse bias* across the power supply. If the power gets hooked up backwards, the diode will be forward biased and will act as a short, keeping the reversed voltage from your circuit.

THE BASICS OF WIRE-WRAPPING

15. In digital circuits, there can be a problem with current demand on the power supply. To filter the noise on the power supply, a 100 μF electrolytic capacitor and 0.1 μF monolithic capacitor is connected from VCC to ground, in parallel with each other, at the entry point of the power supply to the board. These two together will filter both the high and the low frequency noises. Instead of using two capacitors in parallel, you can use a single 20-100 μF tantalum capacitor. Remember that the long lead is the positive one. See Chapter 18 in Volume II for a discussion of decoupling capacitors.

16. To filter the transient current, use a .1 μF monolithic caps for each IC. Place the .1 μF monolithic cap between VCC and ground of each IC. Make sure the leads are as short as possible.

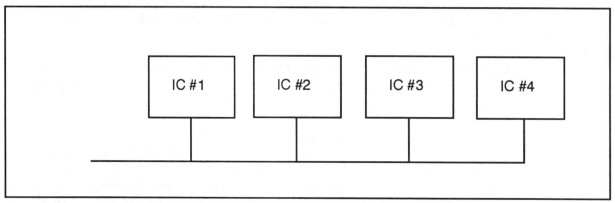

Figure 1. Daisy Chain Connection (not recommended for power lines)

Thank you to Shannon Looper and Greg Boyle for their assistance on this appendix.

Appendix C

DESIGNING YOUR OWN PC BUS EXTENDER

You can buy a PC Bus Extender from a supplier listed in Appendix E of this lab manual or, if you prefer, you can design and wire-wrap your own. This is a guide on how to design and wire-wrap your own PC bus extender.

Note: For further reference, see Chapters 12 and 27 in the textbook and Appendix B of this lab manual.

Materials you will need to make your own PC bus extender include:

Wire-wrap tool
Fully isolated solid 30-AWG wire-wrap wire
Wire-wrap 50-pin header
8-bit PC (PC/XT) interface card(Radio Shack part number 276-1598)
50-conductor cable with 50-pin wire-wrap head (or 25-pin connector cable)
80x86 IBM PC/Compatible computer with a free ISA expansion slot
74LS245, 74LS244 (2), 74LS138

In this guide we assume you are already familiar with wire-wrapping and have good deal of experience in doing so. In Part A we discuss how to buffer the I/O addresses 300 - 31FH. In Part B we show how the I/O addresses can be set to any address using DIP switches. Part C shows how you can include memory addresses.

Part A: Buffering I/O addresses 300 - 31FH

PC BUS Extender for I/O Ports 300 - 31FH

We will discuss designing the bus extender for I/O ports only since free memory space beyond address C0000H varies from PC to PC. As discussed in Chapter 12 of the textbook, the I/O address space of 300H - 31FH is set aside for prototype cards designed by the user. We assume that in your PC, I/O addresses 300 - 31FH are free. Part B discusses a design in which the user can select the I/O address.

Buffering the busses

In designing the PC bus extender, the following points must be noted.

1. The 8-bit data bus must be buffered by a 74LS245.
2. All the A0 - A9 address lines are buffered by the 74LS244.
3. AEN is buffered by 74LS244.
4. IOR and IOW signals are buffered by 74LS244.
5. All the $\overline{OC}$ pins of the 74LS244 are connected to ground, making the 74LS244 tri-state buffers permanently enabled.
6. The IOR signal activates the DIR input of the 74LS245. This ensures that bidirectional data bus buffer is enabled for both read and write. Therefore, when IOR=0 and IOW=1, it is reading data into the PC bus. Conversely, when IOR=1 and IOW=0 the data is being written to an output port. Notice that they can never be high at the same time.
7. The $\overline{OC}$ pin of the 74LS245 is activated by the output of the decoder which decodes only the I/O address range 300H - 31FH and the AEN signal. This ensures that the data bus buffer is enabled only when the I/O address space 300H - 31FH is being accessed.

   Notice that the AEN signal is included in decoding since only the CPU must be in charge when I/O ports are accessed. Note also that AEN=0 when the CPU is in charge.
8. The RESDRV pin is buffered by the 74LS244. Some designs do not use RESDRV; however, it is good practice especially when using the 8255 PPI chip for I/O.
9. All the IC chips on the expansion card are powered by the V_{cc} and ground pins of the expansion slot of the motherboard.
10. Connect all the signals to a connector (header) so that a ribbon cable brings all the signals out of the PC. You can use a 50-pin, 40-pin, or 34-pin header. This allows the extra wires to be used for grounds. If you use a 34- or 40- or 50-pin connector/cable, set aside a ground conductor in between each signal. This reduces the problem of crosstalk.

Figure 1 shows a PC bus extender design for I/O ports 300 - 31FH using a simple 74LS138. You can design your own address decoder using PAL. Implement Figure 1 and get ready for wire-wrapping the circuit.

Wire-wrap the circuit in Figure 1 on a PC/XT add-in card. Many suppliers sell the add-in card. Radio Shack has one with part number 276-1598. Drop by a Radio Shack store and get a catalog showing all the electronics parts they carry. It is a great catalog for any do-it-yourself electronics/computer hobbyist. The fact that there is a Radio Shack on every corner makes it much easier to find and purchase parts. In many electronics parts stores you might find PC/XT expansion cards at an affordable price.

Note: In bringing out the signals (A0 - A9, D0 - D7, IOR, IOW, RSTDRV, and AEN from the 62-pin section of the expansion slot) for the I/O bus extender you must bring out the ground signal as well, but not the Vcc signal. Therefore you need a +5 volt (regulated) power supply to power your circuits on the breadboard. In other words we *do not use the +5V signal from the motherboard to power our own circuit setups on the breadboard. How-*

ever, the ground signal from the expansion slot must be brought out and connected to the ground signal of our circuits on the breadboard. As you have studied in your introductory course on electronics circuits, we can have many power supplies of +5 V, but they all must share a common ground.

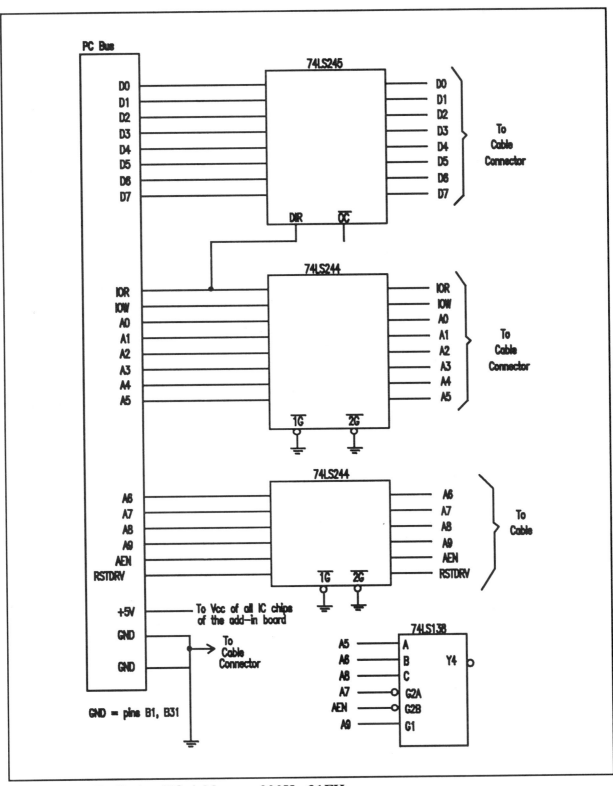

Figure 1. Buffering I/O Addresses 300H - 31FH

Part B: User's selectable I/O address using DIP switches

In the design of Part A, we assumed that the I/O address space 300 - 31FH was free. There are times that some other add-in cards such as PC-based logic analyzers are using a portion of that address space. In such cases, we can design a card where the address space used can be set by DIP switches. Indeed this is standard practice in all commercial products. To allow the selection of I/O addresses, we use a 74LS688 comparator. This is shown in Figure 2. Notice that Figure 2 is the same as Figure 1, except that we have added the comparator and DIP switches for address selection.

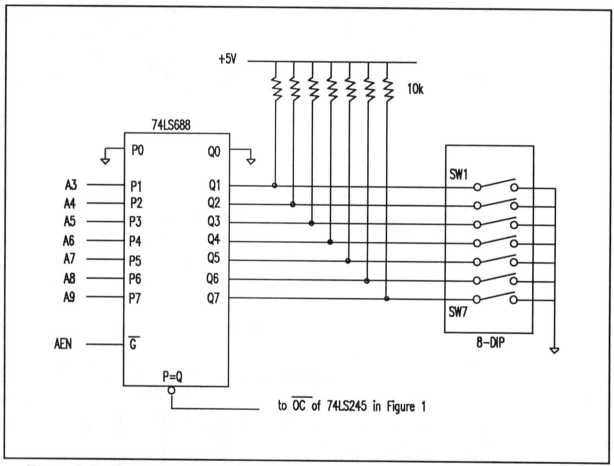

Figure 2. Buffering User-Defined I/O Addresses

Switch Values								Buffered I/O Address
SW8	SW7	SW6	SW5	SW4	SW3	SW2	SW1	300 to 308H
nc	1	1	0	0	0	0	0	

or

SW8	SW7	SW6	SW5	SW4	SW3	SW2	SW1	308 to 30FH
nc	1	1	0	0	0	0	1	

Part C: Memory address buffering

The following is a modification of I/O buffering to include the memory address. This is given only for the sake of understanding and is not recommended for your I/O card.

To include the memory address space we must OR the MEMR signal with IOR. Then the output of the OR gate will activate the DIR pin of the 74LS245. Similarly, for the memory address, the decoder for the memory address space, say D0000 - DFFFF, is also ORed with the I/O address space of 300 - 31FH before it is connected to the G input of the 74LS245. This is shown in Figure 3.

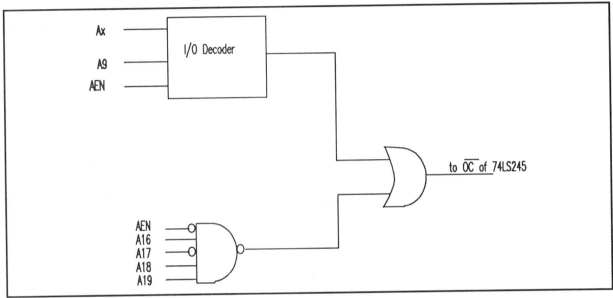

Figure 3. Including Memory Address Buffering

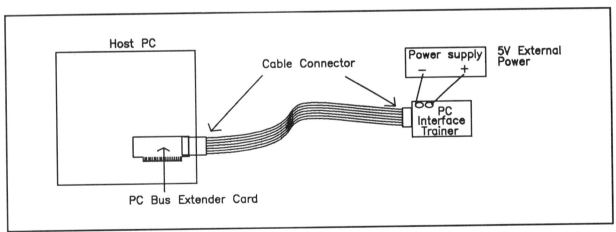

Figure 4. PC Interface Trainer Connections to PC and Power Supply

---

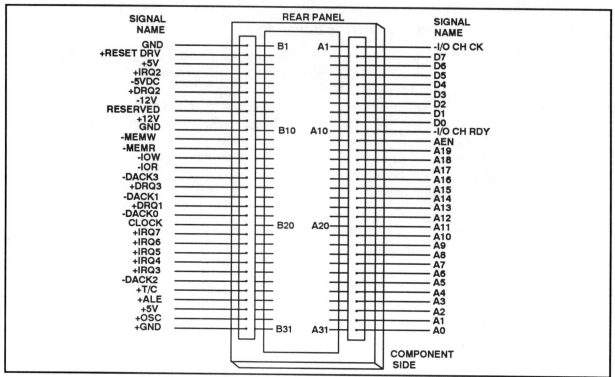

Figure 5. 8088 PC/XT Expansion Slot Detail
(Reprinted by permission from "IBM Technical Reference" c. 1984 by International Business Machines Corporation)

Appendix D

WIRE-WRAPPING A PC INTERFACE TRAINER

Note: For further reference, see Chapters 9 and 12 in the textbook and Appendix B of this lab manual.

Materials you will need to make your own PC bus extender include:
PC Bus Extender accessing the signals of ISA bus
8255 chip
+5 V power source
74LS20, 74LS04, 74LS138, 74LS244
6118 LEDs driver
T1-3/4 (size 5mm) LED, a minimum of 8 of them
8-position DIP switches (SP/ST)
220 ohms resistor (8 of them)
10K ohms resistor (8 of them)

Prior to wire-wrapping this board, you need to study Appendix B in this lab manual and make sure you have some experience in wire-wrapping before you embark on building this circuit. Study Section 12.7 of the textbook and examine the design and interfacing of the PC Interface Trainer. In this appendix we provide a guide on building and wire-wrapping it.

On the wire-wrap perfboard, lay out the chips for the circuit shown in Figure 1. In building this circuit, note the following points.

1. On one side of the perfboard, an edge connector brings in all the signals from the PC bus extender cable.
2. On the opposite side of the perfboard, we mount the +5V and ground binding posts to be connected to an external power supply. At this end, we also connect the decoupling capacitor (10 μF Tantalum) and diode (1N5338) for the entry point of the power as discussed in Appendix B.
3. The ground signal coming from the PC bus extender is connected to the ground signal of the external power supply (that is the ground binding post at the right side of the board).
4. Connect the signals from the PC bus extender edge connector to the appropriate ICs as shown in Figure 1.
5. In some instances, our breadboard circuit needs some of the signals from the PC bus extender. Therefore, we must provide every signal from the PC bus extender to the user. To do that, we connect every signal of the bus extender to a single pin connector and mark each one clearly as shown in Figure 2. You can use what is called a single-pin SIP socket (SIP stands for single inline

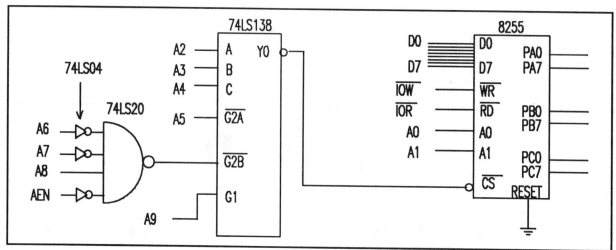

Figure 1. PC Interface Trainer Circuit (I/O Address 300 - 303H)

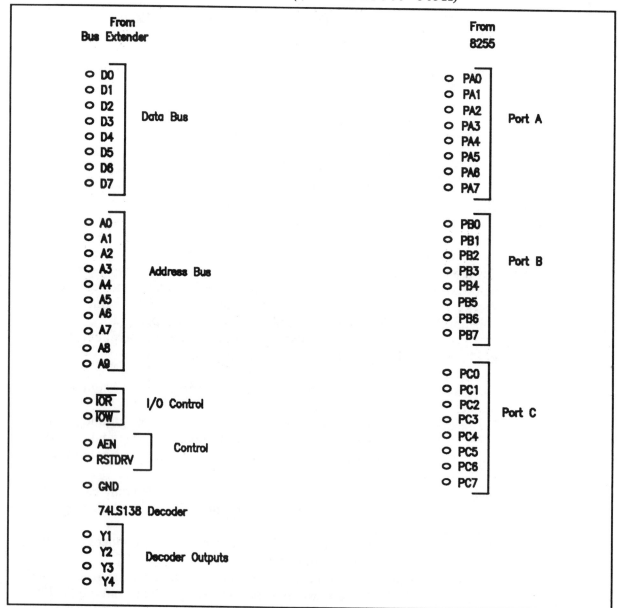

Figure 2. User-Accessible Signals on PC Interface Trainer via a Single-Pin SIP Socket

pin). This way the user can access any of the PC bus extender signals by simply connecting a wire into the hole.

6. All the ports of the 8255 (PA0 - PA7, PB0 - PB7, and PC0 - PC7) are made accessible to the user by connecting each one to a single pin SIP socket. Again, make sure that each bit is clearly marked as shown in Figure 2. You can use the single pin SIP socket. This way the user can connect these signals to his/her breadboard circuit by simply plugging a wire into a hole.

7. Also make Y1 to Y7 of the 74LS138 available by using a single-pin SIP socket. In this way, you will have access to decoders. You will need one for the 8253/54 timer if you are setting up one on your breadboard.

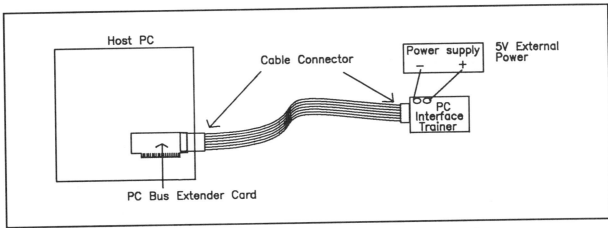

Figure 3. PC Interface Trainer Connections to PC and Power Supply

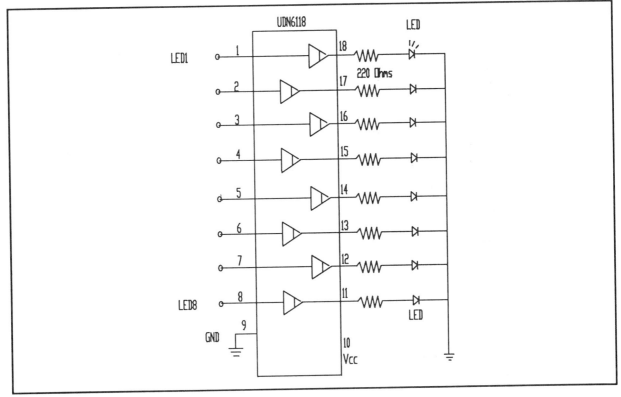

Figure 4. LED Drivers

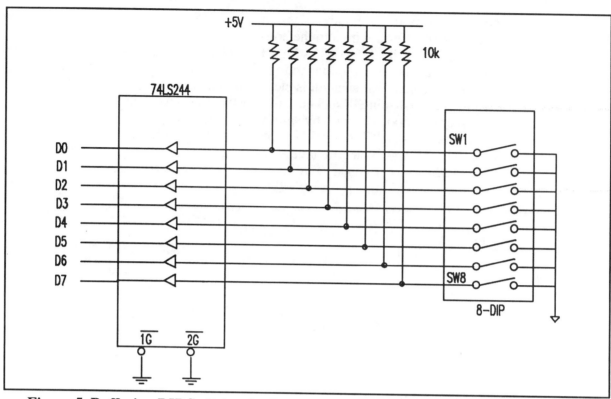

Figure 5. Buffering DIP Switches for Input (the 74LS240 may also be used)

Building LED driver and buffered switches

Although you can use your digital trainer switches and LEDs for testing of the 8255, it is much more convenient to have them available on the PC Interface Trainer itself. The circuits in Figures 4 and 5 show the LEDs and the drivers plus the buffered DIP switches, respectively. It is highly recommended to wire-wrap them on your PC Interface Trainer. Again you must use a single-pin SIP socket to make each LED and switch available to the user. Make sure that they are marked clearly.

After you have wire-wrapped the circuits in Figures 4 and 5, test each LED and switch. In many labs in Sections 2 and 3, whenever we refer to LEDs and switches, we are referring to these two.

Now after connecting your PC Interface Trainer and the LEDs and switches you should test them.

Note: For the LED driver you can use the 74LS04 (or 74LS240) in place of the UDN6118, except you need to make it sink current instead of sourcing it. This is shown as follows.

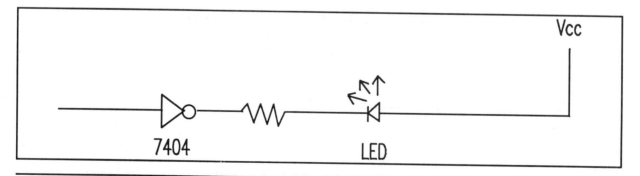

Appendix E

SUPPLIER LIST

The following is a list of suppliers for many of the electronics parts used in this lab manual.

The PC Interface Trainer is available from RSR Electronics.

1. RSR Electronics
    Electronix Express
    365 Blair Road
    Avenel NJ 07001
    Fax: (732) 381-1572
    Mail Order: 1-800-972-2225
    In New Jersey: (732) 381-8020
    www.elexp.com

2. Altex Electronics
    11342 IH-35 North
    San Antonio TX 78233
    Fax: (210) 637-3264
    Mail Order: 1-800-531-5369
    www.altex.com

3. Digi-Key
    1-800-344-4539 (1-800-DIGI-KEY)
    FAX: (218) 681-3380
    www.digikey.com

4. Radio Shack Mail order: 1-800-THE-SHACK

5. JDR Microdevices
    1850 South 10th St.
    San Jose CA 95112-4108
    Sales 1-800-538-5000
    (408) 494-1400
    Fax: 1-800-538-5005
    Fax: (408) 494-1420
    www.jdr.com

6.   Mouser Electronics
       958 N. Main St.
       Mansfield TX 76063
       1-800-346-6873
       www.mouser.com

7.   Jameco Electronic
       1355 Shoreway Road
       Belmont CA 94002-4100
       1-800-831-4242
       (415) 592-8097
       Fax: 1-800-237-6948
       Fax: (415) 592-2503
       www.jameco.com

8.   B. G. Micro
       P. O. Box 280298
       Dallas Tx 75228
       1-800-276-2206 (Orders Only)
       (972) 271-5546
       Fax: (972) 271-2462
       This is an excellent source of LCDs, ICs, keypads, etc.
       www.bgmicro.com

9.   Tanner Electronics
       1301 West Beltline Rd, Suite 119
       Carrollton TX 75006
       (972) 242-8702
       This is an excellent source of LCDs and ICs.

Appendix F

SECTION F.1: 8255 PROGRAMMABLE PERIPHERAL INTERFACE

8255A/8255A-5
PROGRAMMABLE PERIPHERAL INTERFACE

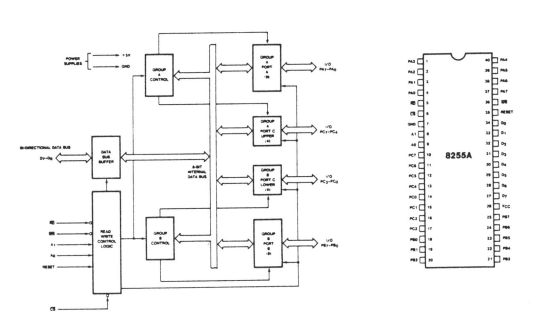

Figure 1. 8255A Block Diagram

Figure 2. Pin Configuration

Courtesy of Intel

8255A FUNCTIONAL DESCRIPTION

General

The 8255A is a programmable peripheral interface (PPI) device designed for use in Intel® microcomputer systems. Its function is that of a general purpose I/O component to interface peripheral equipment to the microcomputer system bus. The functional configuration of the 8255A is programmed by the system software so that normally no external logic is necessary to interface peripheral devices or structures.

Data Bus Buffer

This 3-state bidirectional 8-bit buffer is used to interface the 8255A to the system data bus. Data is transmitted or received by the buffer upon execution of input or output instructions by the CPU. Control words and status information are also transferred through the data bus buffer.

Read/Write and Control Logic

The function of this block is to manage all of the internal and external transfers of both Data and Control or Status words. It accepts inputs from the CPU Address and Control busses and in turn, issues commands to both of the Control Groups.

(CS)

Chip Select. A "low" on this input pin enables the communiction between the 8255A and the CPU.

(RD)

Read. A "low" on this input pin enables the 8255A to send the data or status information to the CPU on the data bus. In essence, it allows the CPU to "read from" the 8255A.

(WR)

Write. A "low" on this input pin enables the CPU to write data or control words into the 8255A.

(A_0 and A_1)

Port Select 0 and Port Select 1. These input signals, in conjunction with the RD and WR inputs, control the selection of one of the three ports or the control word registers. They are normally connected to the least significant bits of the address bus (A_0 and A_1).

8255A BASIC OPERATION

A_1	A_0	$\overline{RD}$	$\overline{WR}$	$\overline{CS}$	INPUT OPERATION (READ)
0	0	0	1	0	PORT A ⇒ DATA BUS
0	1	0	1	0	PORT B ⇒ DATA BUS
1	0	0	1	0	PORT C ⇒ DATA BUS
					OUTPUT OPERATION (WRITE)
0	0	1	0	0	DATA BUS ⇒ PORT A
0	1	1	0	0	DATA BUS ⇒ PORT B
1	0	1	0	0	DATA BUS ⇒ PORT C
1	1	1	0	0	DATA BUS ⇒ CONTROL
					DISABLE FUNCTION
X	X	X	X	1	DATA BUS ⇒ 3-STATE
1	1	0	1	0	ILLEGAL CONDITION
X	X	1	1	0	DATA BUS ⇒ 3-STATE

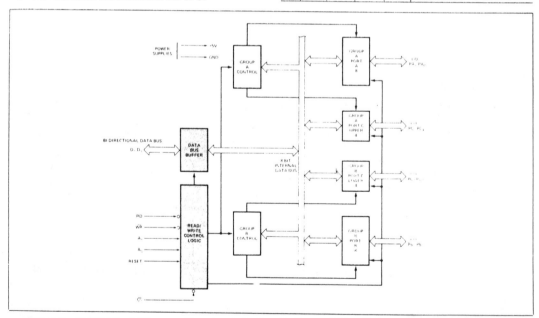

Figure 3. 8255A Block Diagram Showing Data Bus Buffer and Read/Write Control Logic Functions

Courtesy of Intel

(RESET)

Reset. A "high on this input clears the control register and all ports (A, C, C) are set to the input mode.

Group A and Group B Controls

The functional configuration of each port is programmed by the systems software. In essence, the CPU "outputs" a control word to the 8255A. The control word contains information such as "mode", "bit set", "bit reset", etc., that initializes the functional configuration of the 8255A.

Each of the Control blocks (Group A and Group B) accepts "commands" from the Read/Write Control Logic, receives "control words" from the internal data bus and issues the proper commands to its associated ports.

Control Group A — Port A and Port C upper (C7-C4)
Control Group B — Port B and Port C lower (C3-C0)

The Control Word Register can **Only** be written into. No Read operation of the Control Word Register is allowed.

Ports A, B, and C

The 8255A contains three 8-bit ports (A, B, and C). All can be configured in a wide variety of functional characteristics by the system software but each has its own special features or "personality" to further enhance the power and flexibility of the 8255A.

Port A. One 8-bit data output latch/buffer and one 8-bit data input latch.

Port B. One 8-bit data input/output latch/buffer and one 8-bit data input buffer.

Port C. One 8-bit data output latch/buffer and one 8-bit data input buffer (no latch for input). This port can be divided into two 4-bit ports under the mode control. Each 4-bit port contains a 4-bit latch and it can be used for the control signal outputs and status signal inputs in conjunction with ports A and B.

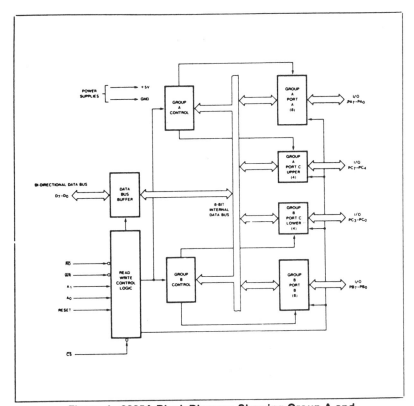

Figure 4. 8225A Block Diagram Showing Group A and Group B Control Functions

PIN CONFIGURATION

PIN NAMES

D₇–D₀	DATA BUS (BI-DIRECTIONAL)
RESET	RESET INPUT
C̄S̄	CHIP SELECT
R̄D̄	READ INPUT
W̄R̄	WRITE INPUT
A0, A1	PORT ADDRESS
PA7-PA0	PORT A (BIT)
PB7-PB0	PORT B (BIT)
PC7-PC0	PORT C (BIT)
Vcc	+5 VOLTS
GND	0 VOLTS

SECTION F.1: 8255 PROGRAMMABLE PERIPHERAL INTERFACE

ABSOLUTE MAXIMUM RATINGS*

Ambient Temperature Under Bias. . . . . 0°C to 70°C
Storage Temperature . . . . −65°C to +150°C
Voltage on Any Pin
   With Respect to Ground. . . . . −0.5V to +7V
Power Dissipation . . . . 1 Watt

NOTICE: Stresses above those listed under "Absolute Maximum Ratings" may cause permanent damage to the device. This is a stress rating only and functional operation of the device at these or any other conditions above those indicated in the operational sections of this specification is not implied. Exposure to absolute maximum rating conditions for extended periods may affect device reliability.

D.C. CHARACTERISTICS   (T_A = 0°C to 70°C, V_{CC} = +5V ± 5%, GND = 0V)

Symbol	Parameter	Min.	Max.	Unit	Test Conditions
V_{IL}	Input Low Voltage	−0.5	0.8	V	
V_{IH}	Input High Voltage	2.0	V_{CC}	V	
V_{OL} (DB)	Output Low Voltage (Data Bus)		0.45	V	I_{OL} = 2.5mA
V_{OL} (PER)	Output Low Voltage (Peripheral Port)		0.45	V	I_{OL} = 1.7mA
V_{OH} (DB)	Output High Voltage (Data Bus)	2.4		V	I_{OH} = −400µA
V_{OH} (PER)	Output High Voltage (Peripheral Port)	2.4		V	I_{OH} = −200µA
I_{DAR}[1]	Darlington Drive Current	−1.0	−4.0	mA	R_{EXT} = 750Ω; V_{EXT} = 1.5V
I_{CC}	Power Supply Current		120	mA	
I_{IL}	Input Load Current		±10	µA	V_{IN} = V_{CC} to 0V
I_{OFL}	Output Float Leakage		±10	µA	V_{OUT} = V_{CC} to 0V

NOTE:
1. Available on any 8 pins from Port B and C.

CAPACITANCE   (T_A = 25°C, V_{CC} = GND = 0V)

Symbol	Parameter	Min.	Typ.	Max.	Unit	Test Conditions
C_{IN}	Input Capacitance			10	pF	fc = 1MHz
$C_{I/O}$	I/O Capacitance			20	pF	Unmeasured pins returned to GND

A.C. CHARACTERISTICS   (T_A = 0°C to 70°C, V_{CC} = +5V ± 5%, GND = 0V)

Bus Parameters

READ

Symbol	Parameter	8255A		8255A-5		Unit
		Min.	Max.	Min.	Max.	
t_{AR}	Address Stable Before READ	0		0		ns
t_{RA}	Address Stable After READ	0		0		ns
t_{RR}	READ Pulse Width	300		300		ns
t_{RD}	Data Valid From READ[1]		250		200	ns
t_{DF}	Data Float After READ	10	150	10	100	ns
t_{RV}	Time Between READs and/or WRITEs	850		850		ns

Courtesy of Intel

A.C. CHARACTERISTICS (Continued)

WRITE

Symbol	Parameter	8255A		8255A-5		Unit
		Min.	Max.	Min.	Max.	
t_{AW}	Address Stable Before WRITE	0		0		ns
t_{WA}	Address Stable After WRITE	20		20		ns
t_{WW}	WRITE Pulse Width	400		300		ns
t_{DW}	Data Valid to WRITE (T.E.)	100		100		ns
t_{WD}	Data Valid After WRITE	30		30		ns

OTHER TIMINGS

Symbol	Parameter	8255A		8255A-5		Unit
		Min.	Max.	Min.	Max.	
t_{WB}	WR = 1 to Output[1]		350		350	ns
t_{IR}	Peripheral Data Before RD	0		0		ns
t_{HR}	Peripheral Data After RD	0		0		ns
t_{AK}	ACK Pulse Width	300		300		ns
t_{ST}	STB Pulse Width	500		500		ns
t_{PS}	Per. Data Before T.E. of STB	0		0		ns
t_{PH}	Per. Data After T.E. of STB	180		180		ns
t_{AD}	ACK = 0 to Output[1]		300		300	ns
t_{KD}	ACK = 1 to Output Float	20	250	20	250	ns
t_{WOB}	WR = 1 to OBF = 0[1]		650		650	ns
t_{AOB}	ACK = 0 to OBF = 1[1]		350		350	ns
t_{SIB}	STB = 0 to IBF = 1[1]		300		300	ns
t_{RIB}	RD = 1 to IBF = 0[1]		300		300	ns
t_{RIT}	RD = 0 to INTR = 0[1]		400		400	ns
t_{SIT}	STB = 1 to INTR = 1[1]		300		300	ns
t_{AIT}	ACK = 1 to INTR = 1[1]		350		350	ns
t_{WIT}	WR = 0 to INTR = 0[1,3]		450		450	ns

NOTES:
1. Test Conditions: 8255A: C_L = 100pF; 8255A-5: C_L = 150pF.
2. Period of Reset pulse must be at least 50μs during or after power on. Subsequent Reset pulse can be 500 ns min.
3. INTR↑ may occur as early as $\overline{WR}$↓.

A.C. TESTING INPUT, OUTPUT WAVEFORM

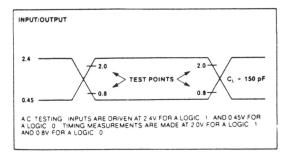

A.C. TESTING LOAD CIRCUIT

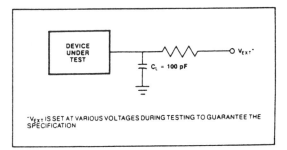

Courtesy of Intel

SECTION F.2: ADC DATA SHEET

National Semiconductor

ADC0801/ADC0802/ADC0803/ADC0804/ADC0805
8-Bit μP Compatible A/D Converters

General Description

The ADC0801, ADC0802, ADC0803, ADC0804 and ADC0805 are CMOS 8-bit successive approximation A/D converters that use a differential potentiometric ladder—similar to the 256R products. These converters are designed to allow operation with the NSC800 and INS8080A derivative control bus with TRI-STATE® output latches directly driving the data bus. These A/Ds appear like memory locations or I/O ports to the microprocessor and no interfacing logic is needed.

Differential analog voltage inputs allow increasing the common-mode rejection and offsetting the analog zero input voltage value. In addition, the voltage reference input can be adjusted to allow encoding any smaller analog voltage span to the full 8 bits of resolution.

Features

■ Compatible with 8080 μP derivatives—no interfacing logic needed - access time - 135 ns
■ Easy interface to all microprocessors, or operates "stand alone"

■ Differential analog voltage inputs
■ Logic inputs and outputs meet both MOS and TTL voltage level specifications
■ Works with 2.5V (LM336) voltage reference
■ On-chip clock generator
■ 0V to 5V analog input voltage range with single 5V supply
■ No zero adjust required
■ 0.3″ standard width 20-pin DIP package
■ 20-pin molded chip carrier or small outline package
■ Operates ratiometrically or with 5 V_{DC}, 2.5 V_{DC}, or analog span adjusted voltage reference

Key Specifications

■ Resolution   8 bits
■ Total error   ± ¼ LSB, ± ½ LSB and ± 1 LSB
■ Conversion time   100 μs

Typical Applications

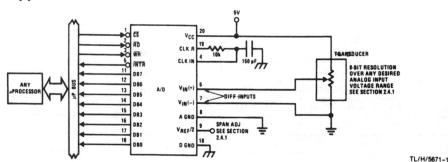

TL/H/5671-1

8080 Interface

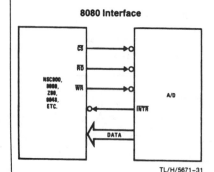

TL/H/5671-31

	Error Specification (Includes Full-Scale, Zero Error, and Non-Linearity)		
Part Number	Full-Scale Adjusted	$V_{REF}/2 = 2.500 V_{DC}$ (No Adjustments)	$V_{REF}/2 =$ No Connection (No Adjustments)
ADC0801	± ¼ LSB		
ADC0802		± ½ LSB	
ADC0803	± ½ LSB		
ADC0804		± 1 LSB	
ADC0805			± 1 LSB

8255A/8255A-5

A.C. CHARACTERISTICS (Continued)

WRITE

Symbol	Parameter	8255A Min.	8255A Max.	8255A-5 Min.	8255A-5 Max.	Unit
t_{AW}	Address Stable Before WRITE	0		0		ns
t_{WA}	Address Stable After WRITE	20		20		ns
t_{WW}	WRITE Pulse Width	400		300		ns
t_{DW}	Data Valid to WRITE (T.E.)	100		100		ns
t_{WD}	Data Valid After WRITE	30		30		ns

OTHER TIMINGS

Symbol	Parameter	8255A Min.	8255A Max.	8255A-5 Min.	8255A-5 Max.	Unit
t_{WB}	WR = 1 to Output[1]		350		350	ns
t_{IR}	Peripheral Data Before RD	0		0		ns
t_{HR}	Peripheral Data After RD	0		0		ns
t_{AK}	ACK Pulse Width	300		300		ns
t_{ST}	STB Pulse Width	500		500		ns
t_{PS}	Per. Data Before T.E. of STB	0		0		ns
t_{PH}	Per. Data After T.E. of STB	180		180		ns
t_{AD}	ACK = 0 to Output[1]		300		300	ns
t_{KD}	ACK = 1 to Output Float	20	250	20	250	ns
t_{WOB}	WR = 1 to OBF = 0[1]		650		650	ns
t_{AOB}	ACK = 0 to OBF = 1[1]		350		350	ns
t_{SIB}	STB = 0 to IBF = 1[1]		300		300	ns
t_{RIB}	RD = 1 to IBF = 0[1]		300		300	ns
t_{RIT}	RD = 0 to INTR = 0[1]		400		400	ns
t_{SIT}	STB = 1 to INTR = 1[1]		300		300	ns
t_{AIT}	ACK = 1 to INTR = 1[1]		350		350	ns
t_{WIT}	WR = 0 to INTR = 0[1,3]		450		450	ns

NOTES:
1. Test Conditions: 8255A: C_L = 100pF; 8255A-5: C_L = 150pF.
2. Period of Reset pulse must be at least 50μs during or after power on. Subsequent Reset pulse can be 500 ns min.
3. INTR↑ may occur as early as $\overline{WR}$↓.

A.C. TESTING INPUT, OUTPUT WAVEFORM

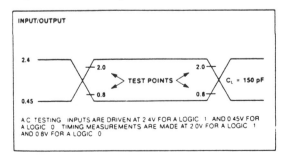

A.C. TESTING LOAD CIRCUIT

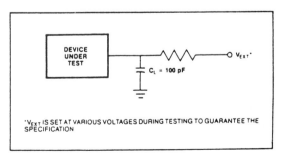

Courtesy of Intel

SECTION F.1: 8255 PROGRAMMABLE PERIPHERAL INTERFACE 245

National Semiconductor

ADC0801/ADC0802/ADC0803/ADC0804/ADC0805
8-Bit μP Compatible A/D Converters

General Description

The ADC0801, ADC0802, ADC0803, ADC0804 and ADC0805 are CMOS 8-bit successive approximation A/D converters that use a differential potentiometric ladder—similar to the 256R products. These converters are designed to allow operation with the NSC800 and INS8080A derivative control bus with TRI-STATE® output latches directly driving the data bus. These A/Ds appear like memory locations or I/O ports to the microprocessor and no interfacing logic is needed.

Differential analog voltage inputs allow increasing the common-mode rejection and offsetting the analog zero input voltage value. In addition, the voltage reference input can be adjusted to allow encoding any smaller analog voltage span to the full 8 bits of resolution.

Features

- Compatible with 8080 μP derivatives—no interfacing logic needed - access time - 135 ns
- Easy interface to all microprocessors, or operates "stand alone"

- Differential analog voltage inputs
- Logic inputs and outputs meet both MOS and TTL voltage level specifications
- Works with 2.5V (LM336) voltage reference
- On-chip clock generator
- 0V to 5V analog input voltage range with single 5V supply
- No zero adjust required
- 0.3″ standard width 20-pin DIP package
- 20-pin molded chip carrier or small outline package
- Operates ratiometrically or with 5 V_{DC}, 2.5 V_{DC}, or analog span adjusted voltage reference

Key Specifications

- Resolution   8 bits
- Total error   ± ¼ LSB, ± ½ LSB and ±1 LSB
- Conversion time   100 μs

Typical Applications

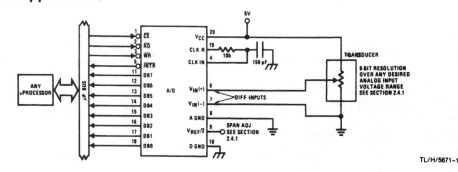

TL/H/5671-1

8080 Interface

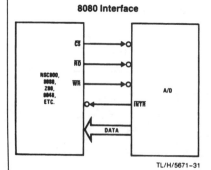

TL/H/5671-31

Part Number	\multicolumn{3}{c}{Error Specification (Includes Full-Scale, Zero Error, and Non-Linearity)}		
	Full-Scale Adjusted	$V_{REF}/2 = 2.500$ V_{DC} (No Adjustments)	$V_{REF}/2 =$ No Connection (No Adjustments)
ADC0801	± ¼ LSB		
ADC0802		± ½ LSB	
ADC0803	± ½ LSB		
ADC0804		±1 LSB	
ADC0805			±1 LSB

Courtesy of National Semiconductor

SECTION F.3: DAC DATA SHEET

 **MOTOROLA**

MC1408
MC1508

Specifications and Applications Information

EIGHT-BIT MULTIPLYING
DIGITAL-TO-ANALOG
CONVERTER

SILICON MONOLITHIC
INTEGRATED CIRCUIT

EIGHT-BIT MULTIPLYING DIGITAL-TO-ANALOG CONVERTER

. . . designed for use where the output current is a linear product of an eight-bit digital word and an analog input voltage.

- Eight-Bit Accuracy Available in Both Temperature Ranges
  Relative Accuracy: ±0.19% Error maximum
  (MC1408L8, MC1408P8, MC1508L8)
- Seven and Six-Bit Accuracy Available with MC1408 Designated by 7 or 6 Suffix after Package Suffix
- Fast Settling Time – 300 ns typical
- Noninverting Digital Inputs are MTTL and CMOS Compatible
- Output Voltage Swing – +0.4 V to -5.0 V
- High-Speed Multiplying Input
  Slew Rate 4.0 mA/μs
- Standard Supply Voltages: +5.0 V and
  -5.0 V to -15 V

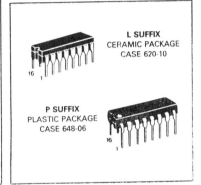

L SUFFIX
CERAMIC PACKAGE
CASE 620-10

P SUFFIX
PLASTIC PACKAGE
CASE 648-06

FIGURE 1 – D-to-A TRANSFER CHARACTERISTICS

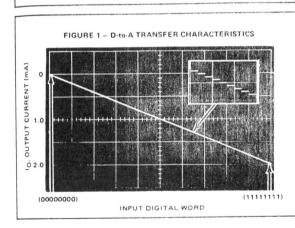

FIGURE 2 – BLOCK DIAGRAM

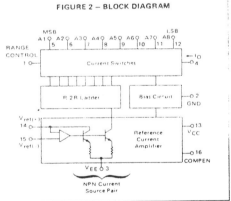

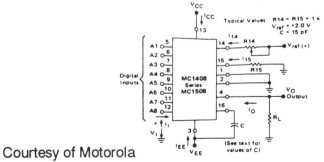

V_I and I_I apply to inputs A1 thru A8

The resistor tied to pin 15 is to temperature compensate the bias current and may not be necessary for all applications.

$$I_O = K \left\{ \frac{A1}{2} + \frac{A2}{4} + \frac{A3}{8} + \frac{A4}{16} + \frac{A5}{32} + \frac{A6}{64} + \frac{A7}{128} + \frac{A8}{256} \right\}$$

where $K \cong \dfrac{V_{ref}}{R14}$

and A_N = "1" if A_N is at high level
A_N = "0" if A_N is at low level

Courtesy of Motorola

SECTION F.4: ULN2003 DRIVER DATA SHEET

 **MOTOROLA**

HIGH VOLTAGE, HIGH CURRENT DARLINGTON TRANSISTOR ARRAYS

The seven NPN Darlington connected transistors in these arrays are well suited for driving lamps, relays, or printer hammers in a variety of industrial and consumer applications. Their high breakdown voltage and internal suppression diodes insure freedom from problems associated with inductive loads. Peak inrush currents to 600 mA permit them to drive incandescent lamps.

The MC1411,B device is a general purpose array for use with DTL, TTL, PMOS, or CMOS Logic. The MC1412,B contains a zener diode and resistor in series with the input to limit input current for use with 14 to 25 Volt PMOS Logic. The MC1413,B with a 2.7 kΩ series input resistor is well suited for systems utilizing a 5 Volt TTL or CMOS Logic. The MC1416,B uses a series 10.5 kΩ resistor and is useful in 8 to 18 Volt MOS systems.

PERIPHERAL DRIVER ARRAYS

SILICON MONOLITHIC INTEGRATED CIRCUITS

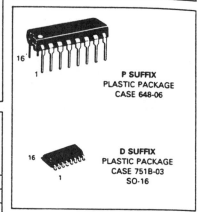

P SUFFIX
PLASTIC PACKAGE
CASE 648-06

D SUFFIX
PLASTIC PACKAGE
CASE 751B-03
SO-16

MAXIMUM RATINGS (T_A = 25°C and rating apply to any one device in the package unless otherwise noted.)

Rating	Symbol	Value	Unit
Output Voltage	V_O	50*	V
Input Voltage (Except MC1411)	V_I	30	V
Collector Current — Continuous	I_C	500	mA
Base Current — Continuous	I_B	25	mA
Operating Ambient Temperature Range MC1411-16 MC1411B-16B	T_A	− 20 to + 85 − 40 to + 85	°C
Storage Temperature Range	T_{stg}	− 55 to + 150	°C
Junction Temperature	T_J	150	°C

Maximum Package Power Dissipation (See Thermal Information Section)
*Higher voltage selection available. See your local representative.

ORDERING INFORMATION

MC1411P (ULN2001A)
MC1412P (ULN2002A)
MC1413P (ULN2003A)
MC1416P (ULN2004A)

MC1411BP (ULQ2001A)
MC1412BP (ULQ2002A)
MC1413BP (ULQ2003A)
MC1416BP (ULQ2004A)

MC1411D
MC1412D } − 20° to
MC1413D   + 85°C
MC1416D

MC1411BD
MC1412BD } − 40° to
MC1413BD   + 85°C
MC1416BD

PIN CONNECTIONS

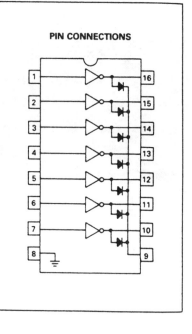

Courtesy of Motorola

LM34/LM34A/LM34C/LM34CA/LM34D

LM34/LM34A/LM34C/LM34CA/LM34D
Precision Fahrenheit Temperature Sensors

General Description

The LM34 series are precision integrated-circuit tempera-ture sensors, whose output voltage is linearly proportional to the Fahrenheit temperature. The LM34 thus has an advan-tage over linear temperature sensors calibrated in degrees Kelvin, as the user is not required to subtract a large con-stant voltage from its output to obtain convenient Fahren-heit scaling. The LM34 does not require any external cali-bration or trimming to provide typical accuracies of ±½°F at room temperature and ±1½°F over a full −50 to +300°F temperature range. Low cost is assured by trimming and calibration at the wafer level. The LM34's low output imped-ance, linear output, and precise inherent calibration make interfacing to readout or control circuitry especially easy. It can be used with single power supplies or with plus and minus supplies. As it draws only 70 μA from its supply, it has very low self-heating, less than 0.2°F in still air. The LM34 is rated to operate over a −50° to +300°F temperature range, while the LM34C is rated for a −40° to +230°F range (0°F with improved accuracy). The LM34 series is available packaged in hermetic TO-46 transistor packages,

while the LM34C is also available in the plastic TO-92 tran-sistor package. The LM34 is a complement to the LM35 (Centigrade) temperature sensor.

Features

- Calibrated directly in degrees Fahrenheit
- Linear +10.0 mV/°F scale factor
- 1.0°F accuracy guaranteed (at +77°F)
- Rated for full −50° to +300°F range
- Suitable for remote applications
- Low cost due to wafer-level trimming
- Operates from 5 to 30 volts
- Less than 70 μA current drain
- Low self-heating, 0.18°F in still air
- Nonlinearity only ±0.5°F typical
- Low-impedance output, 0.4Ω for 1 mA load

Connection Diagrams

TO-46
Metal Can Package*

TL/H/6685–1

*Case is connected to negative pin.
Order Numbers LM34H, LM34AH, LM34CH, LM34CAH or LM34DH
See NS Package Number H03H

TO-92
Plastic Package

BOTTOM VIEW   TL/H/6685–2
Order Number LM34CZ or LM34DZ
See NS Package Number Z03A

Typical Applications

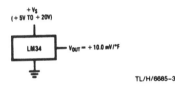

TL/H/6685–3

FIGURE 1. Basic Fahrenheit Temperature Sensor
(+5° to +300°F)

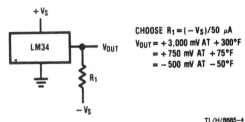

CHOOSE R₁ = (−V_S)/50 μA
V_OUT = +3,000 mV AT +300°F
= +750 mV AT +75°F
= −500 mV AT −50°F

TL/H/6685–4

FIGURE 2. Full-Range Fahrenheit Temperature Sensor

Courtesy of National Semiconductor

LM35/LM35A/LM35C/LM35CA/LM35D
Precision Centigrade Temperature Sensors

General Description

The LM35 series are precision integrated-circuit temperature sensors, whose output voltage is linearly proportional to the Celsius (Centigrade) temperature. The LM35 thus has an advantage over linear temperature sensors calibrated in ° Kelvin, as the user is not required to subtract a large constant voltage from its output to obtain convenient Centigrade scaling. The LM35 does not require any external calibration or trimming to provide typical accuracies of ± 1/4°C at room temperature and ± 3/4°C over a full −55 to +150°C temperature range. Low cost is assured by trimming and calibration at the wafer level. The LM35's low output impedance, linear output, and precise inherent calibration make interfacing to readout or control circuitry especially easy. It can be used with single power supplies, or with plus and minus supplies. As it draws only 60 µA from its supply, it has very low self-heating, less than 0.1°C in still air. The LM35 is rated to operate over a −55° to +150°C temperature range, while the LM35C is rated for a −40° to +110°C range (−10° with improved accuracy). The LM35 series is

available packaged in hermetic TO-46 transistor packages, while the LM35C is also available in the plastic TO-92 transistor package.

Features

- Calibrated directly in ° Celsius (Centigrade)
- Linear + 10.0 mV/°C scale factor
- 0.5°C accuracy guaranteeable (at +25°C)
- Rated for full −55° to +150°C range
- Suitable for remote applications
- Low cost due to wafer-level trimming
- Operates from 4 to 30 volts
- Less than 60 µA current drain
- Low self-heating, 0.08°C in still air
- Nonlinearity only ± 1/4°C typical
- Low impedance output, 0.1 Ω for 1 mA load

Connection Diagrams

TO-46
Metal Can Package*

BOTTOM VIEW

TL/H/5516–1

*Case is connected to negative pin

Order Number LM35H, LM35AH, LM35CH, LM35CAH or LM35DH
See NS Package Number H03H

TO-92
Plastic Package

BOTTOM VIEW

TL/H/5516–2

Order Number LM35CZ or LM35DZ
See NS Package Number Z03A

Typical Applications

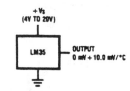

TL/H/5516–3

FIGURE 1. Basic Centigrade Temperature Sensor (+2°C to +150°C)

Choose $R_1 = -V_S/50 \mu A$

$V_{OUT} = +1{,}500$ mV at +150°C
$= +250$ mV at +25°C
$= -550$ mV at −55°C

TL/H/5516–4

FIGURE 2. Full-Range Centigrade Temperature Sensor

Courtesy of National Semiconductor

SECTION F.6: LM741 DATA SHEET

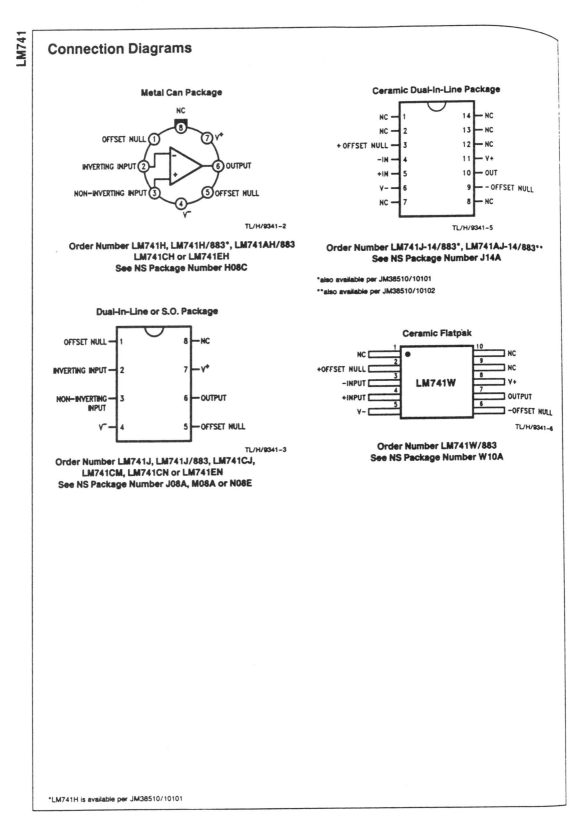

Connection Diagrams

Metal Can Package

Order Number LM741H, LM741H/883*, LM741AH/883
LM741CH or LM741EH
See NS Package Number H08C

Ceramic Dual-In-Line Package

Order Number LM741J-14/883*, LM741AJ-14/883**
See NS Package Number J14A

*also available per JM38510/10101
**also available per JM38510/10102

Dual-In-Line or S.O. Package

Order Number LM741J, LM741J/883, LM741CJ,
LM741CM, LM741CN or LM741EN
See NS Package Number J08A, M08A or N08E

Ceramic Flatpak

LM741W

Order Number LM741W/883
See NS Package Number W10A

*LM741H is available per JM38510/10101

Courtesy of National Semiconductor

SECTION F.7: 74LSXX TTL PIN-OUT

74LS02

SN5402 . . . J PACKAGE
SN54LS02, SN54S02 . . . J OR W PACKAGE
SN7402 . . . N PACKAGE
SN74LS02, SN74S02 . . . D OR N PACKAGE
(TOP VIEW)

1Y [1	14] VCC
1A [2	13] 4Y
1B [3	12] 4B
2Y [4	11] 4A
2A [5	10] 3Y
2B [6	9] 3B
GND [7	8] 3A

74LS04

SN5404 . . . J PACKAGE
SN54LS04, SN54S04 . . . J OR W PACKAGE
SN7404 . . . N PACKAGE
SN74LS04, SN74S04 . . . D OR N PACKAGE
(TOP VIEW)

1A [1	14] VCC
1Y [2	13] 6A
2A [3	12] 6Y
2Y [4	11] 5A
3A [5	10] 5Y
3Y [6	9] 4A
GND [7	8] 4Y

74LS20

SN5420 . . . J PACKAGE
SN54LS20, SN54S20 . . . J OR W PACKAGE
SN7420 . . . N PACKAGE
SN74LS20, SN74S20 . . . D OR N PACKAGE
(TOP VIEW)

1A [1	14] VCC
1B [2	13] 2D
NC [3	12] 2C
1C [4	11] NC
1D [5	10] 2B
1Y [6	9] 2A
GND [7	8] 2Y

74LS30

SN5430 . . . J PACKAGE
SN54LS30, SN54S30 . . . J OR W PACKAGE
SN7430 . . . N PACKAGE
SN74LS30, SN74S30 . . . D OR N PACKAGE
(TOP VIEW)

A [1	14] VCC
B [2	13] NC
C [3	12] H
D [4	11] G
E [5	10] NC
F [6	9] NC
GND [7	8] Y

74LS32

SN5432, SN54LS32, SN54S32 . . . J OR W PACKAGE
SN7432 . . . N PACKAGE
SN74LS32, SN74S32 . . . D OR N PACKAGE
(TOP VIEW)

1A [1	14] VCC
1B [2	13] 4B
1Y [3	12] 4A
2A [4	11] 4Y
2B [5	10] 3B
2Y [6	9] 3A
GND [7	8] 3Y

74LS138

SN54LS138, SN54S138 . . . J OR W PACKAGE
SN74LS138, SN74S138A . . . D OR N PACKAGE
(TOP VIEW)

A [1	16] VCC
B [2	15] Y0
C [3	14] Y1
$\overline{G}$2A [4	13] Y2
$\overline{G}$2B [5	12] Y3
G1 [6	11] Y4
Y7 [7	10] Y5
GND [8	9] Y6

74LS244

(TOP VIEW)

1$\overline{G}$ [1	20] VCC
1A1 [2	19] 2$\overline{G}$/2G*
2Y4 [3	18] 1Y1
1A2 [4	17] 2A4
2Y3 [5	16] 1Y2
1A3 [6	15] 2A3
2Y2 [7	14] 1Y3
1A4 [8	13] 2A2
2Y1 [9	12] 1Y4
GND [10	11] 2A1

74LS245

SN54LS245 . . . J OR W PACKAGE
SN74LS245 . . . DW OR N PACKAGE
(TOP VIEW)

DIR [1	20] VCC
A1 [2	19] $\overline{G}$
A2 [3	18] B1
A3 [4	17] B2
A4 [5	16] B3
A5 [6	15] B4
A6 [7	14] B5
A7 [8	13] B6
A8 [9	12] B7
GND [10	11] B8

74LS373

SN54LS373, SN54LS374, SN54S373,
SN54S374 . . . J OR W PACKAGE
SN74LS373, SN74LS374, SN74S373,
SN74S374 . . . DW OR N PACKAGE
(TOP VIEW)

$\overline{OC}$ [1	20] VCC
1Q [2	19] 8Q
1D [3	18] 8D
2D [4	17] 7D
2Q [5	16] 7Q
3Q [6	15] 6Q
3D [7	14] 6D
4D [8	13] 5D
4Q [9	12] 5Q
GND [10	11] C[1]

74LS688

SN54LS688 . . . J PACKAGE
SN74LS688 . . . DW OR N PACKAGE
(TOP VIEW)

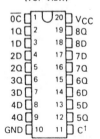

$\overline{G}$ [1	20] VCC
P0 [2	19] P = Q
Q0 [3	18] Q7
P1 [4	17] P7
Q1 [5	16] Q6
P2 [6	15] P6
Q2 [7	14] Q5
P3 [8	13] P5
Q3 [9	12] Q4
GND [10	11] P4

Courtesy of Texas Instruments